HAYWIRE

HAYWIRE

DISCORD IN MAINE'S LOGGING WOODS AND THE UNRAVELING OF AN INDUSTRY

ANDREW EGAN

University of Massachusetts Press
Amherst and Boston

Printed in the United States of America

ISBN 978-1-62534-663-6 (paper); 664-3 (hardcover)

Designed by Deste Roosa
Set in Chaparral Pro and Myriad Pro Condensed
Printed and bound by Books International, Inc.

Cover design by Rebecca Neimark, Twenty-Six Letters
Cover art by Marsden Hartley, *Log Jam, Penobscot Bay, 1940–41.*
Detroit Institute of Arts, USA © Detroit Institute of Arts/Gift of
Robert H. Tannahill/Bridgeman Images

Library of Congress Cataloging-in-Publication Data

Names: Egan, Andrew, author.
Title: Haywire : discord in Maine's logging woods and the unraveling of an industry / Andrew Egan.
Description: Amherst : University of Massachusetts Press, [2022] | Includes bibliographical references and index.
Identifiers: LCCN 2021054626 (print) | LCCN 2021054627 (ebook) | ISBN 9781625346636 (paperback) | ISBN 9781625346643 (hardcover) | ISBN 9781613769416 (ebook) | ISBN 9781613769423 (ebook)
Subjects: LCSH: Logging—Maine. | Lumber trade—Maine. | Loggers—Maine.
Classification: LCC HD9757.M2 E33 2022 (print) | LCC HD9757.M2 (ebook) | DDC 338.4/767409741—dc23/eng/20220125
LC record available at https://lccn.loc.gov/2021054626
LC ebook record available at https://lccn.loc.gov/2021054627

British Library Cataloguing-in-Publication Data
A catalog record for this book is available from the British Library.

A portion of chapter two was previously published as "'The Best Workers They Had'—Le Bûcheron Québécois in Maine's North Woods," in "Canadian Forest History," special issue, *Forestry Chronicle* 90, no. 3 (May/June 2014): 15–18.

To the inherent dignity of woods work
and the earned dignity of those who perform it well.

And to Willa and Etta.

CONTENTS

ACKNOWLEDGMENTS

To those who have come before and, through their work, have contributed in some way to *Haywire*, thank you all: the many observers, researchers, and historians—Kerri Arsenault, Louis-Pierre Bélanger-Ducharme, Robert Bond, Charles Colgan, Louis Hémon, Michael Hillard, Lloyd Irland, Vernon Jensen, Richard Judd, Edward Kendall, Eric Kingsley, Burt Kirkland, Legendre, James McNutt, William Parenteau, Robert Pike, Wayne Reilly, Mark Richard, Félix-Antoine Savard, Howard Schonberger, C. Ross Silversides, David Vail, Graeme Wynn, and many others—who have written about the logging community of the northern forest and whose works are cited herein; the reporters and editors of the *Bangor Daily News*, Mainebiz.biz, *Northern Logger*, *Le Soleil*, *Le Journal de Québec*, *Les Affaires*, and other news sources, on which I have relied for their persistent and objective reportage; to Deryth Taggart, my research associate at the University of Maine, who brought valuable perspective and clarity to the concepts of occupational choice and prestige in the logging community of the northern forest; editors at the University of Massachusetts Press, including Matt Becker, Brian Halley, Rachael DeShano, and Nancy Raynor, and four anonymous peer reviewers, who generously offered their insights into how to improve *Haywire*; and, of course, to the hundreds of loggers from across the northern forest region, including from neighboring Québec, who donated their time and expertise by participating in focused discussions, surveys, and interviews, never reluctant to express their opinions. Finally, thanks to my teachers and role models: Fred Bickford in logging; John O'Brien, Jack Wadsworth, Al Murphy, and Kevin Beattie in forestry; and to my parents, family, and friends in life.

HAYWIRE

Introduction

> Haywire: *A term used contemptuously for poorly equipped loggers. What lies behind this term is the practice of making repairs with haywire.*
>
> —*American Heritage Dictionary of the English Language*, 2006

It all came to a head in the fall of 1975. A century earlier, woods workers in Maine had passed through periods as local suppliers of lumber for building the state's first communities and as farmer-lumberers, dividing their time among the fields during the growing season, the woods in winter, and river drives in the spring. As fields began yielding agricultural commodities rather than simply supporting a subsistence existence among those who worked them, farmers spent more time on the business of farming and less time in the logging woods. As a result, a distinct logging profession began to emerge. Although celebrated as skilled woodsmen and daring river drivers, Maine's loggers would eventually be maligned for being too carefree and too prodigal, for not being enough like the more sober and settled farmer. When technology for using wood fiber instead of increasingly scarce linen and cotton rags to make paper arrived from Europe during the latter part of the nineteenth century, woods workers became logging employees of newly established paper companies. The transition shifted much of their attention from harvesting large sawlogs to smaller pulpwood to feed the paper mills. Maine's north woods would never be the same.

Soon, entire communities—"mill towns"—sprang up from the Maine wilderness. Places with names such as Millinocket, Rumford, Livermore Falls, and Madawaska, built along rivers with such names as Penobscot, Androscoggin, and St. John, thrived on society's turn-of-the-century demand for paper. "The smell of money," the ubiquitous acrid odor that accompanied some paper-manufacturing processes, wandered through town and through forest, a constant reminder of the day-to-day realities to which lives and livelihoods were now inextricably tethered: reliable but debilitating factory and woods work; sylvan countrysides yielding to the fiber demands of forest-consuming mills; the pervasive murmur of clacking millworks and truck traffic; and tributaries of prosperity and pollution coursing through downstream communities. These realities were both comforting and disquieting, life-giving and life-taking—the schizophrenia of Maine's mill towns.

Then, looking for ways to generate even greater profits and to distance themselves from the risks associated with the costs and hazards of felling trees and moving logs from the forest to the mills, companies turned their logging employees loose, creating post–World War II generations of independent logging contractors who owned and operated increasingly mechanized and expensive logging equipment. No longer relying on farm horses for power or the forest products companies for a safety net, Maine's logging community comprised competing, often highly capitalized logging businesses that contracted with mills and forestland owners to harvest and transport raw material from the state's north woods, initiating a stump-to-mill-to-consumer supply chain. However, while valuing a frontier notion of "independence," some loggers soon became disgruntled, unable to successfully transition from logging employee to independent logging business owner.

Eventually many of Maine's newly independent loggers decided to push back. Lacking a clearly defined or shared objective among its leaders, the 1975 labor movement quickly fizzled. But reverberations were felt for decades after, with the increasingly marginalized woods worker, both symbol and symptom of a declining forest industry, striving for greater respect and a more inclusive voice. An industry turned on itself, a logging woods seemingly perpetually distracted by labor discord, unstable markets, and forest politics—a "traditional industry," as one latter-day Maine governor would describe the state's forest products sector, inadvertently signaling an industry in decline, "an industry of the past."[1]

Haywire is about the logging woods of the nation's most forested state at a time of transition in Maine's forests, forest economy, and forest-dependent communities. Drawing on my experience with the region's forest products industry, systematic research, literature review, and media reportage, *Haywire* traces pathways for understanding the challenges facing Maine's logging community and, by extension, the state's forestry sector in the twenty-first century—a period during which Maine's once iconic forest products industry appeared to be unraveling, its statewide relevance fading. It's an industry searching for a path forward in the face of aging mill infrastructure, shifting forest values, and global competition.

Although the focus is on Maine's logging woods, to provide a regional context, *Haywire* places the state of Maine within a northern forest region, an interconnected place comprising New York's Adirondacks; the northern New England states of Maine, New Hampshire, and Vermont; and nearby southern Québec, New Brunswick, and Ontario.[2] Indeed, when it comes to the forest products industry in northern New England, Maine's influence as the region's bellwether—and at one time a major source of wood markets, especially for low-grade logs—can't be ignored.

Fundamentally, *Haywire* is also a story of the realities of logging, of pride and struggle, controversy and conflict, and of trying to make a living in the face of significant political, social, and economic uncertainties. *Haywire* begins by providing historical background about the state's logging woods, a foundation for understanding the evolution of Maine's logging community and forestry sector. Although the first chapter sets the stage by providing a brief overview of the evolution of a northern forest logging profession, this book is not meant to be an epic account of the romanticism of logging in Maine's north woods—of river drives, logging camps, farmer-lumberjacks, and French Canadian loggers—*les bûcherons Canadiens*. Given the cultural significance of the Québécois *bûcherons* in Maine, their contributions to the state's logging woods, and the labor conflict that would inevitably follow, *Haywire* then traces cross-border pathways from the fields and forests of rural Québec to Maine's north woods. A slight diversion from the chronological treatment that precedes and follows it, *Haywire*'s Part 2, while continuing to connect historical threads, lays the groundwork for more completely appreciating the chapters that follow. By highlighting an edited and annotated deposition from an injured northern forest logger, it illustrates the day-to-day realities of working in the most dangerous occupation on land in the United States, along the way providing some background for those who may be unfamiliar with the rudiments of logging jargon and practices.

With a historical context established, the two chapters comprising *Haywire*'s third part take on Maine's post–World War II logging woods of increasingly independent logging contractors and the inevitable conflict around wood prices and Québécois woods labor that followed. Chapter four does much of the heavy lifting in this regard, as it synthesizes and stitches together documentation of the transition from logging employee to independent logger, a necessary prelude to subsequent discussions of discord among Maine's loggers, an increasingly corporatized forest products sector, and antagonism toward Canadian immigrant woods labor. If *Haywire* inadvertently rises to a crescendo, it is owing to the labor action initiated by the Maine Woodsmen's Association (MWA) against the paper companies in October 1975—a relatively short-lived affair of marginal immediate consequence. But despite the movement's shortcomings, *Haywire* posits a persistence in the movement's unrealized, albeit somewhat ambiguous, objectives. Was it about prices that wood-consuming mills paid to loggers that, contractors would contend, barely covered their costs where the mills appeared to be thriving? Was it about resentment toward Québécois loggers, some allegedly hired illegally or shown preferential treatment by being assigned the best logging opportunities? Was it about loggers fighting against their own newfound independence, frustrated by their inability to make it work? Or were there other motivations among the striking woods workers—a lingering inclination toward nativism, an opportunity to strike a blow against a seemingly indifferent forest products industry, or simply a spontaneous hankering for disruption? In the end, even the MWA's organizers couldn't seem to agree. Nevertheless, nearly a half century later, none of those divisive issues has been fully resolved, as Maine's forest industry continues to unravel.

Logging's sometimes petulant history and poor self-image, challenging workplaces, and periodic upheavals lead to questions about the future of the logging workforce explored in *Haywire's* concluding chapters—a historical thread punctuated by disconnects between the harsh realities of logging employment and marginally effective strategies for attracting new workers to an aging industry. The closing chapters update the challenges confronting Maine's forestry sector in the twenty-first century, posing questions along the way about whether their confluence represents a crossroads or a crisis. In so doing, the book concludes by offering commentary derived from the available data and media reports and suggests possible futures for Maine's woods workers, logging woods, and forestry sector based on trends in the state and elsewhere in the region. It brings rather tenuous closure to a story about lessons learned and not learned, of signs of discord and dysfunction carelessly ignored and callously discounted, of an industry that relies on complex ecosystems and a renewable natural resource—the forest—but that has too often undervalued the human and social contexts in which it must function to survive.

Importantly, although logging has at times been casually dismissed as a quaint, rural pastime—"The state of Vermont is supposed to be a picture-taking state," declared one northern forest logger in 2005. "They still want some of us around standing by the highway wearing plaid shirts"—in *Haywire* I keep firmly in mind the words of a Maine logger who declared caustically, "This ain't no way of life, it's my damn job."[3] Given persistent discord in the state's logging woods and the unraveling of its forest economy, the Maine logger's assertion is far from a sure thing.

Part I

Just a Logger

The lumber industry throughout most of its life has been an aggressive, ingenious, unstable migrant. Hard labor, romance of enterprise, and rough, wholesome, and sordid living are found in the chapter in American history which tells of the lumber industry . . . Yet, even though the story is full of exploits of real he-men, it is not a romantic history. It is a history of down-to-earth people struggling and toiling in a harsh environment in order to make a living . . . It is a story of workmen in an oftentimes unnatural, restricted social environment.

—Vernon Jensen, *Lumber and Labor*, 1945

Chapter One
"Almost necessarily poor"

But notwithstanding all the toils of the pursuit, those who once adopt the life of a lumberer seem fond of it. They are in great measure as independent, in their own way, as Indians.

—John McGregor, *Historical and Descriptive Sketches of the Maritime Colonies of British America*, 1828

My day starts about 5 o'clock in the morning. I'm just burned—I'm tired. I'm really dull. I'm about the dullest guy that my wife has ever met. It makes me wonder sometimes, you know, whether it's really all worth it. But I'm stuck with it, because it's a job . . . I'm just a machine. There aren't any rests. When you work in the woods there's no rest because you're there to produce. I'm doing it because I don't know any better and I like the work . . . [but] I wonder how long I can keep up the pace.

—Richard Searls, *Cut and Run*, 1980

This is where logging started and where a logging culture began to develop in the United States and Canada: according to one observer, "the Northeastern region was at one time the leading producing area. It was, moreover, the region in which the pattern of characteristic lumbering activities, practices, and conditions was initially established, and which was reproduced elsewhere when the industry moved on."[1] Aspects of that pattern—including an intergenerational attachment to the profession of logging and the reality of working under challenging environmental conditions while performing a job with low pay and little occupational prestige[2]—have persisted into the twenty-first century.

However, the logging industry didn't simply pull stakes from the Maine woods and the rest of the northern forest and move west, leaving behind cutover forests, struggling mill towns, and unemployed forest workers. Rather, the region's resilient spruce fir, eastern white pine, and hardwood forests recovered, abandoned farm fields reverted to forest, and remnants of the forest products industry remained and at times have thrived, tethered to the often-cyclical demand for lumber, paper, and, more recently, woody biomass for energy.

Those who labored in the logging woods of the northern forest and on the early river drives were often taken for granted, the assumption being that the work was generally unskilled and that there would always be someone desperate or foolish enough to take it on—an attitude that lingers today despite the increasing challenges and complexities associated with woods work and with sustaining a logging business. A quick look back will help us better understand the roots and the persistence of some of these patterns as they have been manifested in Maine's logging community through the early twenty-first century.

FARMER-LUMBERERS

Many of the early loggers in the northern forest region of Maine, New Hampshire, Vermont, and northern New York, as well as nearby New Brunswick and Québec, were "occupational pluralists," farmers opportunistically looking for winter employment for themselves and their workhorses.[3] This was done out of necessity, farmers willing to work in a logging woods that would provide cash income, rather than simply put food on the table, as traditional family-based agriculture had for generations. Working as seasonal loggers, farmers established themselves as key participants in a developing agri-forest economy in order to get ahead, eventually achieving greater self-sufficiency in agriculture. However, it didn't always work out that way for those farmers who were unsuccessful in becoming more occupationally independent, ultimately running the risk of being "ensnared in virtual serfdom to powerful timber barons." As a result, for the less fortunate farmer-lumberers, "the agri-forestry system served as a treadmill for reproducing rural poverty."[4] In addition, much of the early romanticism around logging derived from the contrast between the relatively humdrum subsistence life of the farmer and that of the rogue, daring, more cash-rich logger and river driver. Although the logging woods were tough places to make a living, breaking from the monotony of farm life to an often more exhilarating existence in the woods and in the camps was appealing to many farm-bound young men. As one historian described the flight from field to forest, "The region's farmers who could not make a living from hard-scrabble farms began to desert their land to cut timber, while boys were lured from the plodding labor of the plough by the cash and swagger of the new life."[5]

FIGURE 1. Looking for cash income and work for their horses, farmer-lumberers comprised the workforce of Maine's early logging woods. Image courtesy of Patten Lumbermen's Museum, Patten, Maine.

Since early logging occurred in the fall and winter followed by river driving in the spring, whereas farming activity took place primarily during the summer months, there was a seasonal complementarity between farming and logging.[6] In addition, farmers and loggers possessed similar skill sets, facilitating the transition from agriculture to felling trees and skidding logs during this annual work cycle. "The skills required on the farm were to a large extent the same as those required in the woods," according to one observer. "Skill with an axe and saw and the ability to drive a team of horses were essential in both. Special skills like those of the blacksmith and handyman, and even the amateur veterinarian, were of great value on both the farm and the forest scene."[7] Indeed, it was the early northern forest homesteader and subsistence farmer, not the logger, who was most responsible for land clearing and fallow burning across much of the region's landscape—sometimes quite recklessly—as they struggled to convert virgin forestland into arable farmland and cohesive rural communities as quickly as possible and then to keep the persistent forest at bay.[8] Unfortunately for the image of the region's early woods workers, casual observers of northern forest landscapes often blamed loggers for the careless work of farmers.

In many parts of the region during the early 1800s, farmers and their sons spent some of their relatively slack days between Christmas and Easter at work in the forests.[9] For northern forest loggers from an earlier era, the availability of these famer-lumberjacks and their farm animals, coupled with a reliance on water transportation for winter-cut logs decked on and proximate to frozen lakes, waiting for the thaw that would deliver them to downstream mills, drove much of the logging calendar. For a time, then, farming and logging walked together, defined by a seasonal pattern of interdependence between agriculture and logging and a desire to build rural communities and local subsistence and cash economies. During the mid-1800s, "in northern Maine's isolated frontier society, farming and logging were mutually dependent; settlers provided the labor for both industries . . . The region's small 'backwoods' sawmills of the period operate[d] primarily to meet the lumber demands of local settlers to the region, providing 'a foundation for the frontier society . . . Aroostook County farmers relied upon the lumber industry to funnel cash into the region.'"[10] Indeed, the sawmill often became a center of community activity, serving as a facility to mill grain and card wool.

The lure of logging to boys more acquainted with the slow, more predictable rhythm of farm life was irresistible. "Each year the boy's elder brothers and neighbors would return from the lumber camps, after five months' work, with as much as seventy-five or even a hundred dollars in their pockets, and of course they painted a romantic picture of working and living in the camp," according to one writer.[11] This cash income from logging was an important motivation. Writing about the synergies between logging and agriculture in the Upper St. John Valley, another chronicler of the period wrote: "In many sections of the early nineteenth-century northeast Canadian frontier, agriculture and the lumber industry were closely interconnected . . . Farmers needed the supplementary income from winter work in the lumber camps, as their farms could not sell enough farm products to provide for their needs; and the lumber industry needed the cheap labor the farms represented."[12] A view of the early relationship between farming and logging as it relates to the availability and willingness of farmers to engage in the lumber business was offered by University of Maine historian Richard Judd. Referring to the Aroostook and St. John River valleys of northern Maine, he asserted that "in this last segment of the New England frontier, lumbering and agriculture developed in tandem, mutually dependent, although at times mutually suspicious,"[13] suggesting evolving tensions between the two communities of rural Maine workers.

Other tensions existed, including between loggers and those who employed them, that would presage subsequent conflicts in the logging woods of the northern forest, especially in Maine. Despite prospects of cash income from logging, the region's independent farmers did not always adjust well to their winter reality as logging employees. Maine loggers before World War II were considered "yeoman farmers,"[14] inferring a sense of subordination and servitude. This is consistent with Québécois novelist Félix-Antoine Savard's depiction of Québécois loggers and river drivers in his classic 1937 *roman du terroir* (rural novel), *Menaud, maître-draveur*, in which he described the difficult and hazardous work of *le pauvre draveur* (the poor river driver) in the context of increasing control of Québec's land, forest resources, and forest industry by the English (*les étrangers*):

> *"C'est dur, la drave," dit Alexis . . . "Je ne ferai pas le chien tout mon règne!" . . . Et, sans doute aussi, pensaient-ils à ce qui les attendait: les nuits de glaces, les risques, les misères incroyables . . . sous l'oeil de l'étranger.*
>
> "The river drive is strenuous," said Alexis . . . "I will not make like a dog all of my life!" . . . And, without a doubt, they were thinking of what was awaiting them: the frozen nights, the danger, the incredible hardships . . . under the eye of the stranger.[15]

Savard's depiction of "le pauvre draveur" echoed that of Jeremy Belknap, the historian of early New Hampshire, who in 1792 described the plight of the northern forest logger this way: "The mast trade was formerly confined to England; all white pine trees of certain dimensions being deemed the King's property. The contractors and agents made large fortunes by this traffic; but the laborers who spend their time in the woods and were supplied with provision and clothing for themselves and their families, anticipated their earnings, and were generally kept in a state of poverty and dependence."[16]

Despite the hardships associated with logging and river driving, the attraction of earning cash for their hard work was too much for many of the region's farmers to resist, whereas others may have felt resigned or compelled to work in the woods and on river drives because of the harsh realities associated with farming in the northern forest region, with its bony soils and short growing seasons. However, the connection between farming and logging did not always favor the farmer-lumberer. "In many sections of the early nineteenth-century northeast Canadian frontier, agriculture and the lumber industry were closely connected," according to one observer. But "this resulted in stagnant agriculture and low wages in the lumber industry. There was no incentive to increase agricultural production, and farmers had no choice but to accept the wages and conditions of work offered by lumber companies. The latter, knowing their workers grew most of their food, kept cash wages low and viewed the money as a supplemental rather than a living wage." However, there were other incentives for some farm boys to work in the logging woods and on river drives. Given the region's farming and family culture, not all farmers' sons would have inherited land, especially if the farmstead was small. "Farmers were unwilling to subdivide their farms unless they had a larger-than-average amount of land or several fair-sized holdings . . . Holdings of the average farmer were enough to establish only one child . . . or two at best,"[17] encouraging some farm boys to seek life and livelihood elsewhere.

Eventually, tensions between farming and lumbering grew as agriculture began to move beyond a local, subsistence enterprise after the Civil War, leaving less time for some farmers to engage in logging. Some of the region's farmers also claimed that logging caused them to return to their fields too late in the season to plant and that their draft animals were often too worn out to work in the fields after a winter of toil in the logging woods. Soon, the two professions began to diverge and develop separate personalities, as farming evolved from a subsistence lifestyle to an increasingly commodity-driven, cash enterprise. In essence, "commercialized farming had loosened the ties between lumbering and agriculture," and the days of the farmer-lumberer were numbered. Farmers simply became less dependent on logging for cash to supplement their subsistence livelihoods. In addition, even in the mid- to late 1800s, logging was considered something of a risky business given losses associated with cyclic lumber markets and poor years for in-woods operations. Indeed, the so-called

Long Depression of the mid-1870s punctuated a period of turmoil in the lumber industry. With lumber markets in decline, farming and logging were further separated as farmers looked to diversify their crops beyond what would have normally serviced local consumer-woodsmen.[18]

But given the undeniable economic importance of the timber trade in the early to mid-nineteenth century, why were criticisms of the lumber industry and those who engaged in it so pervasive, whereas farming was spared? One argument goes that in parts of the northern forest in the early nineteenth century, agriculture was a more reliable foundation for society than was the timber trade. Farming was also considered a more rewarding occupation than logging, value-based assertions that were likely connected to a view of loggers as being more itinerant and their profession more seasonal and less stable. "Farmers were virtuous, lumberers were depraved," according to one historian . "Of all the bases for public wealth, farming was considered least liable to fluctuate; it was a stable activity, rooted in the land, not a subservient enterprise dependent upon external markets."[19] The contrast between agriculture and logging was similar in Québec's logging woods:

> *Le travail agricole est jugé clairement moins dur physiquement (la vie est plus longue), le fermier est plus libre et autonome, la ferme lui appartient et assure un revenu plus régulier, sinon lui permet de subsister, et le travailleur y demeure avec sa famille . . . Cela s'explique du fait que la plupart des travailleurs forestiers n'ont pas pu vivre de leur terre et sont justement devenus bûcherons pour cette raison.*
>
> Farmwork was clearly considered less physically strenuous (life was longer), the farmer more independent and autonomous, the farm belongs to him and assures a more regular return, at least allowing him to subsist, and the worker resides with his family . . . That explains why most forest workers have not been able to live off their land and as a result have become loggers for that reason.[20]

In his decidedly shallow attempt to compare farm labor with woods labor during the early twentieth century, academic Burt Kirkland wrote in the prestigious *Journal of Forestry* in 1920: "We have only to compare the average conditions in agriculture with those in forest industry to see how much is missed in the latter as regards pride of workmanship. On the farms we have millions of men planning and carrying out plans for improvement of the industry. Forest industry will produce like results whenever forests are handled for continuous production instead of destruction."[21] While this view certainly held true in some areas of Maine and the rest of the northern forest, it ignored the dependence of the forest industry on sustainably managed forests and the recovery of the northern forest from earlier community-building land clearing and careless fallow burning by famers.[22]

Contributing to logging's image were often poorly cited historical threads that appeared to persist, despite data that might suggest a far different narrative about life and livelihood in the logging woods of the northern forest. Early indictments of loggers and condemnations of the work that they performed were considered by some to be generally inaccurate, "often erected upon the most slender of foundations," relying on "accounts of individual indebtedness or by tales of the springtime debauchery of an itinerant raftsman; similarly, justifications of the farmer's life were found in the success and happiness of a single settler." The same observer of the northern forest logger added that "twentieth-century judgments of New Brunswick's early nineteenth-century lumbering industry were harsh. More contemporary criticisms of the "rising crusade against the forest . . . were frequently exaggerated," describing the "literature" that framed logging and loggers in derogatory language as "the product of visitors to and the elite of the province."[23]

Indeed, it was not uncommon for early logging to be erroneously blamed for the destruction of the northern forest. As described in the book *Adirondack Hard Times*,[24] New York State's northern forest loggers were consistently demonized by casual observers and privileged-class historians for the land clearing and fallow burning that was being performed by farmers, despite evidence that could be found in state reports and newspaper accounts from the period that would suggest otherwise. For example, Alfred Donaldson, the New York City banker turned historian, offered this about logging in his 1921 two-volume *History of the Adirondacks*: "The march of the lumberman was like that of an invading army—they attacked and destroyed the outposts first, and only gradually made their way to the inner citadel. They did damage, because they lumbered carelessly, with no concern for the future. Their worst sin was the fire menace that they left behind, and which caused incalculable destruction."[25] While researching his book, Donaldson would have had access to official state reports of the period and indeed selectively cited some of them in his work. The 1898 Third Annual Report of the Commissioners of Fisheries, Game and Forests of the State of New York, for example, made the following emphatic statement: "The lumbermen cause no fires, despite the common impression to the contrary. During the winter, the time when they are in camp, the snow prevents forest fires."[26] According to the New York Fisheries, Game, and Forest Commission in its 1899 report: "The most frequent cause of woodland fires in our State are the small fires started by farmers for the purpose of burning brush, logs, and stumps, in order to clear some piece of land. These are known locally as fallow fires, and the operation is generally alluded to as burning a foller." Further, absolving loggers, the report continued: "It is the farmer, not the lumberman, who has destroyed so many thousands of acres of timber land. The lumberman takes only a few trees per acre of some merchantable species; the farmer in his operations destroys the entire forest."[27] Nevertheless, loggers of the northern forest remained targets of often undeserved criticism, which formed a historical basis for a profession suffering from a persistent image problem.

"COWBOYS" OF THE NORTHERN FOREST

As agriculture in the northern forest region matured beyond its roots as a local subsistence enterprise, farmers began to realize cash income from their labor, and logging began to be seen as a distinct occupation. Soon, fewer of the region's loggers were farmer-lumberers, and their legend as independent, prodigal adventurers, no longer anchored to the relative stability of the family farm, began to develop. Describing the evolving distinction between early farmer-lumberers and professional loggers, one historian asserted that "the earliest loggers in New England were, of course, Yankees, and they came from the 'poverty' farms that bore most of the population. There was no labor-saving machinery, everything was done by 'main strength and ignorance.'" Commenting on the untethered life of the new breed of logger, he added: "But the professional loggers, especially the unmarried ones, were very often drifters. Urged on by an incurable wanderlust, by a taste for novelty, they enjoyed going from river to river, from camp to camp—partly, no doubt, to show off their prowess and match themselves against other redoubtable members of the craft."[28]

Despite logging's early romanticization, history often treated the logger and his profession poorly, and especially when compared to the more stable profession of farming, attitudes about loggers and logging did not improve much over the decades. According to the English author and traveler Edward Augustus Kendall, writing in 1809 after a journey through the northeastern United States: "Maine is covered with wood; wood, in all its forms, is marketable; and to the settlers is therefore equivalent to furs or fish; and the settlers have consequently degenerated, not into hunters nor fishermen, but into *lumberers*" (Kendall's emphasis), asserting that "he that lives on the bounty of nature, lives always in poverty." Comparing loggers with the more virtuous "husbandman, whose toil is always for the future," Kendall asserted that "the lumberer toils only for the moment that is passing and provides for that moment only by preying on the future one: what nature has planted he enjoys, but he plants nothing for himself," likely unaware of the resilience of the northern forest and its ability to regenerate itself. Nevertheless, Kendall kept on with his denunciation of loggers, making a rather oblique connection between the nature of logging work and the lack of virtue of those who perform it: "His habits in the forest, and his voyages for the sale of his lumber, all break up the system of persevering industry, and substitute one of alternate toil and indolence, hardship and debauch; and, in this alternation, indolence and debauch will inevitably be indulged in the greatest possible proportion. Nor is this all; the lumberer is nurtured in dishonesty, not less than in idleness and intemperance; and he is nurtured, not only in habits of idleness, intemperance and dishonesty, but in the habits of an outlaw and desperado."[29] Describing the early northern forest logger of southern Ontario during the early 1800s, when the British navy began to send "gangs of axemen" into the woods specifically to cut lumber and masts, thereby creating a separate occupation, one chronicler of the period asserted that "townsmen began to view the new breed of woodsmen with some apprehension. They were, at the very least, a colorful bunch, shaggy of hair and beard, dressed out in red and blue and green jerseys, with knitted sashes about their waists, and red and blue and green tuques on their heads."[30] The stereotypical logger, as well as the reputation that accompanied him, continued to evolve, as loggers were compared with rough-and-tumble workers from other fields. "The lumberjack, together with the sailor, has been described as reckless, social and generous," according to one historian. "He has also been likened to the cowboy, the droves of logs corresponding to the herds of cattle, the log mark the brand. The contemporary attitude of the lumberjack varied in tone. Several of the earlier commentators were hostile; later travelers were generally admiring; native writers were irritatingly enthusiastic." Apparently referencing Kendall's earlier wanderings through Maine's north woods, he wrote that "at the beginning of the century an English visitor spoke of the Maine lumberman in harsh terms because he believed that the logger was misusing the resources of the State. It is evident that he felt that Maine ought to be a grazing and farming state,"[31] again demonstrating a developing bias against logging and in favor of agriculture, despite the region's growing reliance on forests and loggers to meet the demand for forest products and to help establish and diversify an early northern forest economy.

A broad and not always very flattering stereotype of the northern forest woodsman thus evolved, becoming entrenched in regional lore and history. "Those who are lumbermen are almost necessarily poor. Their course of life seduces them to prodigality, thoughtlessness of future wants, profaneness, irreligion, immoderate drinking and other ruinous habits," according to one writer in 1821. "The farmers of New-England [*sic*] have never willingly resided among people of such character."[32] Joining a growing chorus of anti-logger sentiment, another early observer of the northern forest woods worker asserted seven years later that the "moral character of loggers, with few exceptions, is dishonest and worthless. I believe that there are few people in the world, on whose promises less faith can be placed, than on those of the lumberer . . . Lumberer is considered synonymous with a character of spendthrift habits and villainous, vagabond principles. After selling and delivering up their rafts, they pass some weeks in idle indulgence; drinking, smoking, and dashing off, in a long coat, flashy waistcoat and trowsers [*sic*], Wellington or hessian boots, a handkerchief of many colors around the neck."[33]

Further contrasting farmer-lumberers and the new breed of full-time, year-round woods worker, one historian described the logging profession this way: "Others in the work force were lumberjacks plain and simple, carrying on after the spring and summer drive with rafting and driving operations, as camp watchmen, sometimes as axemen on timber cruising parties, and at just about any task that might be open to them, either in woods or sawmill." Making a distinction between early "farmer-lumberjacks" and "drifter-lumberjacks" of the 1920s and 1930s, he described the latter as "highly mobile" and "nomadic" with "no farm to go back to . . . It is probably this breed of lumberjack who has contributed most to a widely held image of the rollicking, hard-drinking, womanizing, hell-raising guy. When they saw him coming, proprietors were inclined to close down the bars while anxious mothers pulled their daughters off the street behind closed doors."[34] According to another historian, "With a few exceptions, men in the industry, both employers and employees, have been extreme individualists," which has "helped give lumbering a picturesque history marked by exploits of the traditional 'he-man.'"[35]

The image of the prodigal, cocksure logger and its associated impacts on logging's occupational prestige was amplified in the early twentieth century. Recounting a situation in Bangor, Maine, in 1914 when hundreds of destitute immigrant loggers—"These aren't tramps or bums," the Bangor police chief was quoted as saying. "Of course, there may be a few bums among them"—were stranded in town after coming down from the woods when the winter logging season had ended early because of mild weather and river drivers were waiting for their season to begin as snow and ice was just beginning to melt, a reporter noted that "there probably isn't a driver in the lot. That is a kind of work requiring not only strength and endurance but the agility of an acrobat and the admirable daring that has made the name and fame of Hancock Street and the Hamden Road echo through the land. It takes the Bangor boys—the Irish and Yankees, the Old Town Indians and the little and wiry French Canadians—to work the white water and handle the cantdog and setting pole. Germany, Sweden, Russia and Finland would be merely in the way on the East Branch."[36]

Other writers have attempted to characterize the psychosocial character of the logger, generally with little or no substantiation of their assertions. Quoting from the *Report of President's Mediation Commission to the President of the United States* in the fall of 1917, for example, Kirkland offered the following about the American logger in the prestigious *Journal of Forestry*: "Partly the rough pioneer character of the industry, but largely the failure to create a healthy social environment has resulted in the migratory, drifting character of (woods) workers. Ninety percent of those in the camps are described by one of the wisest students of the problem, not inaccurately, as 'womanless, voteless, jobless.' The fact is that about

FIGURE 2. The early history of Maine's logging woods has often been described in romantic terms: daring men taming the state's north woods. Day-to-day realities of isolating logging camp life, harsh working conditions, and injury and death in the logging woods and on the river drives told a different story. Image courtesy of Patten Lumbermen's Museum, Patten, Maine.

ninety percent of them are unmarried. Their work is most intermittent . . . They are—or rather have been made—disintegrating forces in society." Discussing the untethered existence of the logger during that period, Kirkland added this analysis: "It is thus evident that one of the most outstanding evils of the transient character of destructive lumbering on labor is the denial of the family life. The average laborer entering this industry inevitably ends his family line. The repression of his normal sex life also in most instances introduces vagaries of various kinds into his character which when combined with a continued sense of injustice, due to the uncertain character of his job, and its attendant conditions, makes his mind fertile ground for freak economic and political theory, to such an extent that masses of men subjected to such conditions are shaking the foundation of organized society."[37] As with the observations made by Kendall more than a century earlier, it is unclear from his writing what Kirkland meant by "freak economic and political theory" or how loggers were "shaking the foundation of organized society."

A logger's separation from his family and family life was considered by others to be harmful, even immoral. Writing about the isolated working and living conditions of northern forest loggers from New Brunswick, one historian observed: "Others denounced the harmful social effects of forest employment. In their view, lumbering drew people away from home and substituted alternate bouts of toil and indolence, hardship and debauchery, for consistent industry and 'the good social effects consequent upon being surrounded by women.'" Comparing loggers of the era to gypsies, he noted: "Once the winter was over, idleness, drunkenness, and gambling were the lumberers' past-times [*sic*]." A logger's life cycle was reduced to spending "their winter in the woods and their summer lounging about in the towns."[38] Clearly, the image of the northern forest logger was not always consistent with the often highly technical reality of their work. According to one observer of the logging community in 1920, "Forestry requires intelligent, skilled labor in the woods." Presaging future challenges related to recruiting new workers into the logging workforce, he asserted that "to obtain and hold good men for woods work it is essential to make employment as permanent as possible, and so to adjust conditions that the workers will be interested in and satisfied with their jobs, transforming the present 'blanket stiff' type of labor into that of the decent, self-respecting, home-making citizen. The woodsman must first be contented with his job if we are successfully to begin the practice of forestry."[39]

Reality or mythology, whatever one's take on loggers during this period, the eventual disappearance of these often-colorful northern forest woodsmen was cause for some regret, leading one writer to state wistfully in 1932 that "the reckless men, the legendary American lumberjacks, passionate, vigorous, impetuous daredevils . . . are now almost a vanished

FIGURE 3. Maine woodsmen often suffered from a poor image, despite the skills required to safely and efficiently fell trees and transport logs from the woods to downstream mills. Image courtesy of Patten Lumbermen's Museum, Patten, Maine.

race . . . The ubiquitous double-breasted red-flannel undershirt of the old-time lumberjack has vanished."[40] Stated perhaps more emphatically in 1967 by another historian, "I know that the present-day woods-laborer is economically far better off than the impoverished woodmen who worked for Van Dyke sixty years ago, and I am glad of it, but as Ernest Martin Hopkins, the late president of Dartmouth College, wrote to me shortly before he died, I know too that he is a damned sight less interesting."[41] As we'll see later, some of Maine's latter-day loggers might argue about whether they are better off than their 1960s predecessors and would likely not care about whether they might be considered "less interesting" by the president of an elite university or anyone else.

As the era of the colorful northern forest logger began to fade, a romanticism around old-time loggers and logging began to take hold. "So the old-fashioned, red-shirted, two-fisted, calk-booted woodsman of former eras has disappeared, and the tar-paper flaps forlornly on the roofs of the old log camps," lamented one logging historian.[42] In addition to increases in the amount of in-woods mechanization, there were other, more pervasive changes occurring in the region's logging woods as the culture of logging was beginning to shift. "But of later years, with government doles, unions, and other factors, woods help has become scarcer and harder to get, and jobbers have become more honest," declared the historian.[43] About a decade later, another chronicler of logging in the northern forest stated that, after World War II, the "spirit and atmosphere" of the logging woods of the northern forest were also changing. "Much of the color—some called it 'glamour'—was fading fast."[44]

Hobbled by dangerous work and a harsh work environment, and "almost necessarily poor," many Maine loggers eventually looked elsewhere for safer, more lucrative year-round employment. Woods help often became scarce, especially in the most remote forest enclaves of the state, forcing the industry to look further north, especially to neighboring Québec, for men who were desperate or foolish enough to take on Maine's logging woods.

Despite often romantic depictions of early logging in Maine—of choppers and axemen, raftsmen and river drivers pitted against the north woods in a heroic contest of man against nature—day-to-day logging woods realities told a different story. "It is a history of down-to-earth people struggling and toiling in a harsh environment in order to make a living," according to one writer. "It is a story of workmen in an oftentimes unnatural, restricted social environment," a reference to both isolating logging camp life and remote logging workplaces.[45]

At first, farmers looking for supplementary income and winter work for themselves and for their workhorses comprised the region's logging workforce. But the seasonal farmer-lumberer cycle began to break down as agriculture moved from a preoccupation with subsistence and survival to a year-round occupation that brought cash income to the farm family. As a result, a more distinct logging profession, untethered to the seasonality of farmwork, emerged. At the same time, a less flattering image of the northern forest logger began to develop. Ignoring the skills, knowledge, and discipline needed to work safely and efficiently in the woods, erstwhile experts characterized loggers as homeless, penniless, and womanless. Others described them as reckless drifters and troublemakers, "disintegrating forces" who shook "the foundation of organized society." Making matters worse for the woodsman's image, casual observers of the northern forest landscape often blamed loggers for the large land clearings and burned-over forests left behind by farmers who struggled to keep a persistent forest at bay. Logging also suffered from comparisons with agriculture—"farmers were virtuous, lumberers were depraved"—a reflection of the stability of farming and the prodigality of woods work and life in the logging camps.

Eventually, recruiting new loggers to this punishing work and its isolated workplaces and marginal living conditions became more challenging. Maine's logging workforce began to depend more and more on immigrant labor, especially in the state's more remote north woods.

Chapter Two
"The best workers they had"

The most insulting thing to the (Maine) woodsmen was that the Canadians were harder workers.

—Michael Hillard, interview with Don Fontaine, July 15, 2004

As we've seen, the forest industry of Maine has traditionally relied on woods workers from Québec to complement the region's logging workforce, especially in the state's more remote northern outposts. The historical concentration of harvesting in Maine in winter derived, in part, from the seasonal availability of labor and workhorses from farms on both sides of the United States–Canadian border, a practice that continued for several decades.[1] According to a 1986 Maine Department of Conservation report, Canadian workers have crossed the border to work in the Maine woods since the 1800s: "They actually lived closer than Maine residents to the 'unorganized territories' of Maine where most logging took place and have always been described as good workers for relatively low pay."[2] Given the proximity of rural settlements in eastern Québec to remote logging chances in northern Maine, these "farmer-lumberjacks" were often easier to recruit and retain than were domestic woods workers.[3]

However, despite a rich history of Québécois forest workers in Maine's logging woods, the relationship between the logging communities on both sides of the border has not always been amicable. A combination of economic, political, and demographic factors posed significant challenges to the reliance of northern forest landowners and forest products businesses on Québécois woods labor. Tensions between American loggers and the large Maine paper and land management companies that control the forest products industry in the northern part of the state have been recurrent in the state's logging woods, as has a persistent sense of resentment and nativism toward Québécois loggers, who have often been perceived by some of their American counterparts as taking their jobs and by others as harder workers.[4] Controversy and conflict related to immigrant labor in Maine's logging woods were just around the corner, as tensions between American loggers, their Canadian counterparts and the paper companies who hire them to cut wood were coming to a head.

C'ÉTAIT BÛCHE, BÛCHE, BÛCHE

Canadian workers from the province of Québec began working in Maine's woods during the mid-nineteenth century, becoming the skilled *draveurs* (river drivers) and *bûcherons* (loggers) of the region's northern forest. Their reputation as industrious and daring woods workers was legendary. French Canadian workers from the province of Québec began working in Maine's woods during the mid-nineteenth century. Portrayed in historical descriptions of the region and celebrated in Québécois cultural and artistic expression, they were generally described in positive, respectful terms, as reflected in this refrain from Roger Miron's "Hommage aux bûcherons":

Oh é il est infatigable
Bon travailleur et gai luron

He is tireless
Good worker and cheerful lad

Louis Hémon, a French novelist who spent a period of his life working as a farmhand in Québec, offered the following description of woods work and Québécois woods workers and river drivers, based on actual characters and events, in his classic *roman du terroir*, *Maria Chapdelaine*, written in 1913:

> *Les chantiers, la drave, ces sont les deux chapitres principaux de la grande industrie du bois, qui pour les hommes de la province de Québec est plus importante encore que celle de la terre. D'octobre à avril les haches travaillent sans répit et les forts chevaux traînent les billots sur la neige jusqu'aux berges des rivières glacées; puis, le printemps venu, les piles de bois s'écroulent l'une après l'autre dans l'eau neuve et commencent leur longue navigation hasardeuse à travers les rapides. Et à tous les coudes des rivières, à toutes les chutes, partout où les innombrables*

> *billots bloquent et s'amoncellent, il faut encore le concours des draveurs forts et adroits, habitués à la besogne périlleuse, pour courir sur les troncs demisubmergés, rompre les barrages, aider tout le jour avec la hache et la gaffe à la marche heureuse des pans de forêt qui descendent.*
>
> The lumber camps, the drive, these are the two principal aspects of the great wood industry, that for the men of the Province of Québec is more important than that of [working] the land. From October to April the axes work without rest and the draft horses drag the logs over the snow to the banks of the frozen rivers; then the spring comes, and the piles of wood melt one after another into the rising waters and begin their long hazardous journey through the rapids. And at every bend, at every falls, everywhere where the vast quantity of logs jam and pile, it is necessary with the help of strong and skillful river drivers, practiced at the perilous work of running across the half-submerged logs, to break the log jams, to assist with ax and pike pole the free descent of this mass of floating forest.

However, descriptions of the life of the Québécois logger have not always been as epic, and the life of the early French Canadian logger was far from romantic, *le bûcheron* not always a "cheerful lad." Hémon also described *la misère* of the Québécois river driver this way:

> *On n'avait pas de tentes non plus pendant l'été: rien que des abris en branches de sapin qu'on se faisait soi-même, et du matin à la nuit c'était bûche, bûche, bûche, mangé par les mouches et dans la meme journée trempé de pluie et rôti de soleil.*
>
> We didn't have tents during the summer: just shelters made of spruce branches that we made ourselves, and from morning until night it was chop, chop, chop, eaten by the flies and in the same day soaked by the rain and roasted by the sun.[5]

Given the ethnic diversity of early loggers in the northern forest, some historians couldn't resist their own, often biased, and undocumented characterizations of the region's logging workforce, including *le bûcheron Québécois*. One such raconteur was Alfred Donaldson, who in 1921 wrote the two-volume *History of the Adirondacks*. "The French Canadians as a rule made the best log-drivers and became the most cunning at the tricks of their trade. They seemed naturally endowed with the agility, recklessness, and immunity to exposure that must combine to make the expert. They have always predominated as a race in the lumbering operations in these woods," Donaldson asserted while housebound by tuberculosis in the Adirondack town of Saranac Lake. However, Donaldson's statements contradict the observations of a Yale graduate student, who, some years earlier in his 1902 master's thesis titled "The Lumbering Industry in Maine," wrote: "It is not strange that the best men are Indians and Irishmen. The soft voiced, quiet French Canadian is not keen for this work. When we do find him at it, we invariably find a strong strain of Indian in his blood. On the whole it may be said that the Penobscot Indian makes the best river driver. The big, athletic Irishman, however, is more daring and less careful of his life," broad assertions based on ethnicity that were left unsupported in either his narrative or his bibliography.[6] At the same time, as it became more common for the French language to echo throughout Maine's logging woods, there was little doubt that nativism toward French Canadians was beginning to develop, including in the state's northern forest.

"I WILL NOT SPEAK FRENCH IN SCHOOL"

According to one chronicler of the Acadian experience in Maine's mill towns, "From about 1840 until 1930, almost a million French-speaking Canadians immigrated to the United States, with the largest percentage going to New England to work in its paper and textile mills."[7] The Census Bureau estimated that in 1920, Franco-Americans made up 33 percent of Maine's immigrant population. But many Maine Yankees resented the growing influence of Acadian culture and their general willingness to perform strenuous work in the mills and in the forests. Primarily targeting Acadians living in Aroostook County, a law passed by the Maine State Legislature in 1919 required public schools to teach in English, and teachers were required to punish students who spoke French. Incredibly, the law remained in effect until 1969.[8]

Not all Quebeckers were content with their lot south of the border, especially those who found themselves unwittingly working in Maine's remote north woods. Years earlier, under pressure from the state's influential lumber interests, Chapter 7 of Public Laws of Maine for 1907, widely known as the 1907 Maine peonage law, was passed.[9] It prevented malcontent woods workers, many of whom were Quebeckers and other immigrants who were apparently tricked into taking jobs as loggers by employment agents in Bangor, Lewiston, and Boston, from abandoning the state's logging woods in search of more appealing work elsewhere.[10] Agents were paid bounties of a dollar a head to recruit woods workers from among often unsuspecting immigrants who were driven into Maine's northern forest and dropped off along the road, leaving them with little choice but to walk miles to the nearest logging camp.[11] Described by the New England Historical Society as "near-slaves,"[12] hundreds of these "recruits" were sent to jail under the 1907 peonage law for refusing to become part of the north wood's logging workforce and live in primitive and often unhealthy logging camp conditions.

There were even darker days ahead for the state's immigrant workers. In the 1920s, Maine and much of the rest of the northern forest region saw the reemergence of the Ku Klux Klan. With over one hundred thousand members in the United States, during that period Maine had the largest KKK representation outside the South. New England's post–World War I nativist, antiunion, and anti-Catholic KKK movement primarily targeted Irish immigrants and Franco-Americans. "The KKK considered French Catholics subservient, a dark-skinned race that stole jobs, defiled American culture, and spoke in pagan tongues (Latin masses, French)," according to a writer from one of the state's mill towns.[13] The state of Maine and some of the mills aligned themselves and shared culpability with the KKK for racist policies and attitudes toward the newcomers.

Maine's remote north woods were not immune from the KKK's nativist attitudes toward immigrants from Québec. In 1924, described as "the Ku Klux Klan's most active year in the state of Maine," the so-called Battle of Greenville pitted some forty members of Maine's burgeoning KKK against union organizers and French American loggers whom they were recruiting to join the Industrial Workers of the World (IWW), or Wobblies. The KKK opposed efforts by the Wobblies to unionize loggers within the Lumber Workers' Industrial Union in the small rural town of Greenville in Maine's Piscataquis County, an effort that was also under way in the logging communities of the northern forest states of New Hampshire, Vermont, and New York. The KKK found support among some local lumber companies that didn't want to see an organized logging workforce that would demand higher pay and better working and living conditions. If the union organizers resisted, the KKK threatened to remove them forcibly. But the IWW recruiters were undeterred and continued to organize the region's woodsmen, despite the arrest and conviction of three Wobblies leaders who were accused of "conspiracy" after trying to organize a "boycott against four Greenville establishments for discriminating against its members."[14] The IWW leader proclaimed, "We are going to stick, and if the Klan starts anything, the IWW will finish it." Referring to the Great Northern Paper Company and others as "slave drivers," the IWW declared defiantly that "the day of the old logging camp and the lumberjacks is about over with."[15]

Once confronted, the KKK backed down. The "battle" saw no violence, with the Wobblies returning to the lumber camps to recruit more loggers and the vastly outnumbered KKK retreating to its dark corner. Ultimately, however, the union's attempts to recruit Greenville's French Catholic loggers collapsed under the weight of pressure from mill owners and the town that had been enflamed by the violent and nativist rhetoric of the KKK.[16]

Presaging the Battle of Greenville and subsequent labor tensions in Maine's logging woods, and criticizing both the role that forest industry played in marginalizing its forest workers and the tactics of the IWW, Ferdinand Silcox, chief of the U.S. Forest Service from 1933 to 1939, wrote in his article titled "Forestry and Labor," published in the *Journal of Forestry* in 1920: "Perhaps no industry in the country has had and is continuing to have more 'labor trouble' than the lumber industry. The I.W.W. organization has flourished in the lumber camps, and their destructive type of radicalism is in many cases the result of the failure of forestry to lay the foundation for permanent communities, with home life, and because of the archaically feudalistic conception of the industry generally as to the right of the employees to organize and share with capital and management a voice in the councils of the industry—particularly concerning those immediately vital things that affect directly the workers' interests . . . Is the lumber industry going to adopt democratic industrial relations policies voluntarily, or is it going to challenge the democratic forces of the country to see that it does?"[17] The KKK's membership in Maine declined sharply after the Greenville affair and other attempts to marginalize the state's Catholic immigrant population, and by 1930 there were fewer than 300 Klansmen claiming Maine their home. As asserted by Mark Richard in *Not a Catholic Nation*, the staunch resistance to the KKK by Franco-American immigrants, including at the Battle of Greenville, was largely responsible for racist movement's ultimate demise in Maine.[18] Its central nativist theme, however, lingered among some Mainers and, it has been suggested, was an underlying impulse for subsequent discord in the state's logging woods.

LE BÛCHERON QUÉBÉCOIS

As we've seen, combining the demands of agriculture with a harsh existence as loggers and river drivers was particularly important in sustaining rural family and community economies in Québec. Importantly, logging helped establish and support a cash economy in rural Québec, the seasonal tempo of farming life often in rhythm with that of woods work. As described by Camille Legendre in his 2005 book, *Le travailleur forestier Québécois* (The Québécois forestry worker):

> *Jusqu'a la fin des années 1940, l'industrie d'exploitation forestière sera caractérisée par sa complémentarité avec l'agriculture . . . La main-d'oeuvre est donc recrutée parmi les fermiers et leurs fils, qui trouvent dans le travail en forêt un supplément financier particulièrement utile et apprécié . . . La principale raison d'être des jobbeurs est donc leur capacité de mobiliser la main-d'oeuvre nécessaire pour cette activité saisonnière année après année et de la maintenir au travail.*
>
> Until the end of the 1940s, the logging industry was characterized by its relationship with agriculture . . . The workforce was recruited from among farmers and their sons, who found in forest-related work especially useful and appreciated supplementary income . . . The main reason for being logging contractors was their ability to mobilize the workforce needed for this seasonal work year after year and to keep it working.[19]

This phenomenon continued until the postwar period, when there was a migration from rural areas to the cities, in part owing to industrial expansion, resulting in chronic instability in the province's logging workforce. Later, in the 1960s, significant sociocultural shifts toward a greater liberalization of Québécois society—manifested in Québec's *la révolution tranquille*[20]—would liberate many Quebeckers from feeling obligated by the Catholic Church to maintain their traditional agricultural lifestyles, opening doors for young people to migrate to urban areas to pursue alternative careers away from the province's plowed fields and working forests.

Although the rich history of the Québécois logger in northern Maine provides a backdrop for understanding current and future conditions related to logging in the region and the tensions that would soon follow in the state's logging woods, other, more immediate questions related to occupational choice and familial attachment to logging among Québécois loggers emerged, calling into question the overall stability of the region's logging workforce. A 1960 study found that among Québécois loggers, the most common reasons to choose logging as a profession were the shortage of available work in the parish; lack of skills to do other work; and a lack of education needed to obtain work in the city.[21] Interviews of Québécois loggers living in logging camps in Maine conducted a few years later by the Public Affairs Research Center (PARC) also suggested a sense of resignation to logging among interviewees. Responding to the question, "Why did you take this job?," 31 percent said that they had insufficient education for other work, and 21 percent indicated that they could not find another job nearby. Only 15 percent reported that it was the best-paying job available, while 12 percent said that they logged because they had always done this work. The study further described the northern Maine logger in a way that suggests little sense of occupational identity or choice: "[The logger is] a man with minimal education (by our standards) who has no skills other than his own strength and who sees the logger occupation as the only occupation where he can make a good living (by his standards). His attitudes imply that he would prefer his own children to have a better lot than his, that he has resigned himself to the woods as his only hope, and that he is willing to accept the conditions of the occupation without complaint as long as the money is there."[22] Educational attainment and resignation to their occupation aside, woods workers from Québec, either bonded to an American employer or working in the United States under a visa, were thought by some to be superior in logging skills and work ethic to native Mainers, no doubt raising the hackles of some hardworking Maine loggers. A 1960 review of woods labor in Maine observed that "it was the feeling of one [forestry] company official . . . that the bonded and visa labor were the best workers they had." Defending the use of Canadian labor in Maine's north woods, the author concluded: "[Canadian loggers] are not given preference as some people claim, although they are much better workers on the whole . . . The bonded laborers cause very little trouble to the companies and jobbers; and the companies feel that since they are faced with an acute shortage of domestic labor, the practice of bonding labor is an effective way of meeting the problem at this time."[23] Some years later, Maine loggers would have something to say about the author's assertions.

"A RESERVE ARMY"

The availability of French Canadians to work in the northern forest—especially in Maine's industrial north woods—has been attributed to several factors, including a lack of job opportunities in Québec; a historical lack of adequate education opportunities in rural Québec that could qualify many Quebeckers for alternative employment; easy access to Maine's most northern logging sites; and the presence of many French Canadian Maine citizens, providing opportunity to develop a strong sense of community in the state's rural northern counties. Many Québécois residents were "inclined to seek timber harvesting jobs because they come from rural farm backgrounds where woods work is learned at a young age."[24]

In addition, provincial economic policy generally failed to create wage employment for poor farmers in Québec, compelling them to seek logging employment in remote regions of Maine.[25] By the 1930s, some Québécois moved to Maine to find new logging opportunities, having converted to agriculture or otherwise exploited many of the forests of southern Québec:

> *Les colons ont tôt fait de déboiser leur terre et les lots avoisinants, de façon erratique, pour combler la hausse de la demande*

> *et de profiter des prix élevés du bois . . . Ils sont bientôt obliges d'aller bûcher à l'extérieur, en particulier dans l'État du Maine.*
>
> The settlers [in Québec] had earlier cleared their land and neighboring lots, in an erratic fashion, to meet heightened demand and to take advantage of higher prices for wood . . . They were soon obliged to log elsewhere, particularly in the state of Maine.[26]

An immigrant labor agreement reached in 1940 between the United States and Canada and put into effect the following year solidified the relationship between the two countries and attracted Canadians—mostly Québécois—to the logging woods of the northern forest. While immigrant visa laborers had the same privileges as U.S. citizens (except for the right to vote), bonded labor had to be requisitioned by a U.S. employer through a government-controlled application process. The "bond" was then required to work for the requisitioning employer for a predetermined base rate of pay and return to Canada when work is completed. However, a prospective employer must attempt to recruit and show preference for domestic labor when it was available. Put simply, bonded woodsmen were described as "Canadian nationals who are permitted to enter the United States at the solicitation of employers following a procedure which certifies that no domestic labor is available to perform the type of work for which they are being sought."[27] As described in 1965 by Leo Thibodeau, a manager for Great Northern Paper Company in Bangor, the "bonding program " required "the return of each [bonded] woodsman to Canada upon termination of employment. From 1940 on, this agreement worked very well," providing the company with a stable logging workforce.[28]

However, a decade or so later Québécois woods workers soon realized that it was relatively easy to obtain immigrant visas, encouraging some would-be bonded loggers to find employment in other sectors of the New England economy—presumably where the work was perceived as being more lucrative, less strenuous, and safer—thereby abandoning woods work. "Personally, I've not been happy with the visa situation because it depleted our labor supply," according to Thibodeau. "Only one of every five visa holders continues to work in the woods." There were other reasons besides the harsh and seasonal nature of woods work for Québécois visa workers to move on. "These people, like our [U.S.] citizen woodsmen before them, want better education for their children and more modern conveniences, and these cannot be achieved by working six to nine months a year" in the woods, wrote Thibodeau.[29]

By World War II, loggers from southern Québec began to work in larger numbers in the Maine woods. In the mid-1960s it was estimated that 75 percent of the logging workforce in northern Maine was from Québec, according to the PARC study. In addition, socialized medicine and subsidized credit for the purchase of logging equipment may have combined to encourage Québécois loggers to tolerate greater risks to personal safety and less lucrative logging opportunities than may have been acceptable to their Maine counterparts.[30] "Between 1956 and 1970, the state certified 1,500 to 5,000 'bonded' Canadians each year to cut in situations where (the industrial landowners claimed) Americans were not willing to work," according to one source. "Though firm data are non-existent, it appears that 'bonds' cut 20 per cent to 40 per cent of Maine timber."[31] Nevertheless, observing a decline in the number of bonded Canadians in northern Maine's logging woods from the mid-1950s through the mid-1970s, a 1978 article titled "Status and Outlook of Bonded Canadian Woodworkers in Maine" confidently predicted that "the end of the bonded Canadian logger's era in Maine is in sight."[32]

But not all was well with the Québécois logger working in Maine's north woods, some of whom complained about being discriminated against for supervisory positions in the state's logging woods. A 1974 bulletin from the Yale University School of Forestry on employment stability in the logging industry suggested that loggers were often at an educational disadvantage in the competition for more skilled jobs, placing constraints on an individual's job options and potential for a greater sense of occupational choice.[33] Indeed, loggers in eastern Canada have historically had the lowest education levels of major Canadian census groups, with between 79 and 87 percent of loggers in the eastern provinces of New Brunswick, Newfoundland, Ontario, and Québec possessing an elementary education or less, according to a 1966 study.[34] Comparatively low levels of education among Québécois loggers have persisted, albeit not as emphatically as in previous decades. Writing in 1978 of the "basic problems of Québec today," one Canadian observer of the province's logging sector asserted that "the fact of illiteracy helps to explain why supervisory and management positions usually had to be filled by outsiders. The problem lay in the educational system that simply did not qualify the young Québécois for such advancement."[35]

The lack of employment opportunities in Québec described in earlier decades persisted into the 1970s, encouraging more Québécois loggers to seek employment in Maine's north woods, and the demand for woods workers from Canada appears to have been related to high quit rates among American loggers.[36] The assertion at the time was that American workers didn't want to work in the state's more remote north woods. "Most of those Canadians who are in the Maine woods are up on the border in northwestern Maine where Americans don't want to go . . . I don't know really what would happen if we couldn't bring in some [Canadian loggers]," according to a Maine forest industry representative in 1980.[37] (At the turn of the twenty-first century, little appeared to have changed regarding employment opportunities in the region of Québec proximate to Maine's border counties. In 2001, the Québec economic regions that adjoin Maine—Estrie, Chaudière-Appalaches, and Bas-Saint-Laurent—had unemployment rates of 6.5 percent, 7.3 percent, and 11.0 percent, respectively.[38] In the same year, unemployment rates in Maine's five border counties with Québec ranged from 4.4 percent [Oxford County] to 7.6 percent [Somerset County], according to the Maine Department of Labor.[39])

Immigrant Québécois loggers came to represent a "reserve army" of often impoverished and seasonally underemployed Québécois farmers.[40] Québécois loggers were reportedly also willing to work in regions of Maine considered by most of the state's citizens to be so remote that there have been suggestions that raising wages alone would not attract enough domestic workers to the logging industry to meet labor demands there. The attraction of perhaps better pay and more lucrative logging chances in Maine may have contributed to a logging labor shortage in Québec. Noting a "drastic reduction" of qualified loggers in Québec from 1967 to 1971, a 1973 study of Québec's woodlands labor shortage recommended efforts to improve the image of Québécois loggers, as well as their working conditions and wages.[41]

In the end, Maine's industrial forest complex was generally ambivalent about whether loggers were Canadian or American or whether they spoke

French or English. In addition to an evolving nativism toward Québécois woods workers, the hazardous work environment and often unhealthy living conditions posed ever-present threats to life and limb for those performing the most dangerous job in the country.

Maine has traditionally relied on woods workers from Québec to complement the region's logging workforce, especially in the state's more remote north woods. Over time, a folklore evolved around the Québécois logger and river driver, often feeding off conflicting stereotypes. By the mid-1960s it was estimated that 75 percent of the logging workforce in northern Maine was from Québec. Many of these loggers were bonded workers, described by one source as "Canadian nationals who are permitted to enter the United States at the solicitation of employers following a procedure which certifies that no domestic labor is available to perform the type of work for which they are being sought." They generally traveled across the border each week to cut and skid wood, spending weeknights in logging camps and weekends back home in Québec.

The work was strenuous and camp life was isolating. Studies in the 1960s suggested that loggers from the province took on work in Maine's north woods primarily because of a lack of employment opportunity in Québec and a lack of skills and education for other employment. In general, Québécois loggers who worked in Maine expressed a sense of resignation to their occupation. Despite the challenges, one mill representative described Canadian "bonds" as "the best workers they had." Soon, tensions between resentful Maine loggers and their counterparts from north of the border reverberated through Maine's logging woods. The state of Maine was complicit in the discord, having passed laws decades earlier that were aimed at the forced assimilation of immigrants from Québec.

In at least one instance, local Maine governments and wood-consuming mills sided with the Ku Klux Klan in its attempt to control Catholic, French-speaking Canadian woods workers, who were viewed as "a dark-skinned race that stole jobs, defiled American culture, and spoke in pagan tongues." Nativist attitudes toward Maine's Québécois woods workers would continue to foment discord, playing a role in the ongoing unraveling of the state's forestry sector. However, other threats menaced Maine's northern forest, as the risk of injury and death hung over those who worked in the state's logging woods.

Part II

Maine's Logging Woods

Apart from the atrocious living conditions and lack of normal emotional and psychological outlets, there was the fact that death was always ready to claim its victims with falling branches and trees, rolling and pitching logs, churning rapids encountered on the log drives, and whirring saws and speeding, powerful machinery in the mills.

—Vernon Jensen, *Lumber and Labor*, 1945

Chapter Three

Too Close to the Stump

Premature old age, and shortness of days, form the inevitable fate of a lumberer. Should he even save a little money, which is very seldom the case, and be enabled for the last few years of life to exist without incessant labor, he becomes the victim of rheumatisms and all the miseries of a broken constitution.

—John McGregor, *Historical and Descriptive Sketches of the Maritime Colonies of British America*, 1828

Evolving historical narratives about logging and loggers aside, woodsmen of the northern forest were generally more preoccupied with the difficult day-to-day work of producing raw material to meet the needs of a growing forest products industry, while surviving the most dangerous work in the country. According to the Bureau of Labor Statistics, logging has consistently been the most dangerous occupation on land in the United States. Despite advances in logging mechanization, logger safety training, and a 1995 Occupational Safety and Health Administration logging standard, fatal logging accidents still occur at alarming rates. In 2016, for example, logging experienced a rate of nearly 136 fatal accidents per 100,000 woods workers, with deaths generally attributable to falling objects and moving woods equipment, such as rubber-tired skidders and log loaders. The Bureau of Labor Statistics ranked logging the most dangerous job in the United States in 2018, with 97.6 fatal injuries per 100,000 full-time workers, a rate that was 28 times higher than the all-worker rate of 3.5 fatalities per 100,000 full-time workers. Fisheries was a distant second that year, with 77.4 fatal injuries per 100,000 workers. Although mechanized felling has reduced logging accidents caused by falling trees and branches, logging remains extremely hazardous, with most fatalities occurring close to the stump.[1]

TRAGEDY IN THE WOODLANDS PARISH

The following obituary is from a 1953 issue of the *Northeastern Logger* (formerly the *Lumber Camp News* and now the monthly *Northern Logger* trade journal), an issue that included ads featuring Lombard's "Powerhouse" chain saw, the "Featherweight" IEL chain saw, and the McCulloch Model 4–30, which weighed in at "only 30 pounds" (latter-day production chainsaws weigh about half that weight), just as the chainsaw was being introduced into the logging woods of the northern forest:

> *William Francis Gonyea, 28, of Tupper Lake, New York, was killed instantly on September 15, 1953, when he was struck in the head by a huge dead limb, dislodged as he was felling a hardwood tree.*[2]

Fatalities such as these were not aberrations in the northern forest. Some months later, for example, the same column reported the following logging woods accidents:

> *Bath, NH. A falling log instantly killed Harold E. Ainsworth, 52, local lumberman, while he was engaged in logging operations.*
> *Levis Aucoin, 50, a woodsman employed on a hardwood logging job, was fatally injured when he was struck by a huge falling limb.*
> *Northville, NY. A tree that fell the wrong way crushed a lumberjack to death recently. Floyd Purdell, 40, was cutting down the tree in Sacandaga, NY.*[3]

Although the most precariously exposed to risk of injury, the timber faller, or chopper, was not the only member of a logging outfit who was exposed to danger. Another logging-related death in the same article was attributed to a self-inflicted gunshot wound by a camp cook. The title of the regularly occurring section of the magazine was "Deaths in the Woodlands Parish," suggesting a period when there may have been a greater sense of community and connectivity among loggers, fostered by living together in the same camps and remote timber-dependent settlements, as well as the strong intergenerational and familial connections of the region's early logging workforce. "Injuries were common and the entire camp would stop when the misery whistle blew six blasts, indicating an injury," according to one account of injuries among loggers holed up in a western Canadian logging camp for the cutting season. "Hats would come off when the whistle blew seven times and the camp would have to bury one of their own. A falling snag or a widow maker could easily and indiscriminately take out a buckwheater or an old timer."[4]

Even before the advent of the chain saw, wheeled skidders, and other mechanized equipment in the 1950s, the logging woods of the northern forest presented considerable danger to woods workers. An early account of what was awaiting loggers was offered by New York State forester William Fox in 1902:

> The life of the pioneer woodsmen was always beset with dangers peculiar to their work. Early town records make frequent mention of fatal accidents which befell them. It is remarkable how often the first death in a settlement was of some man killed by a falling tree, of one who was crushed by a load of logs or killed in his sawmill. Not only the pioneers, but their successors, have contributed to the same death roll every year. These causes are various: A heavy limb falls, broken by the wind; a tree "lodges," and, springing back from the stump, kills the axeman; a load of logs "shoves" the team [of draft animals] down some steep grade in the road, and the driver is thrown underneath or dashed against a roadside tree; a tier of logs starts suddenly; a jam on the log drive breaks without warning; a man while fighting a forest fire finds his retreat cut off; another disappears in the current of the spring flood, and in the mills men fall upon the saws.[5]

Clearly, not all logging accidents were fatal, but results were often no less gruesome. Again, according to Fox, describing a logging-related accident in western New York, "It is written in the records of the town of Middlebury, Wyoming County, that—In May, 1817, Artemus Shattuck went into the woods to chop. While cutting off a log that had been partially split open, his foot was caught in the crack, and he hung for a long time suspended by his foot and partly supported by one hand. Despairing of receiving aid, he finally unjoined his ankle with his pocketknife, made a crutch of a crooked stick, and started for the house." In 2020, over two hundred years later, there was this headline from the *Bangor Daily News*: "Logger pinned to tree cuts himself free with chainsaw."[6]

Trees that are lodged, or "hung up," in other trees could be particularly dangerous if a chopper tried to cut the supporting tree, an unsafe but not uncommon practice. As reported by the *Bangor Register* in 1828, long before the introduction of the chain saw: "As Mr. David Inmann of Howland was, on the 29th of January, at the mouth of the Piscataquis, felling a tree upon which another tree was lodged, the lodged tree, as the other tree started, slid down its trunk, struck his left shoulder, broke all his ribs on the left side—broke his thigh bones just above the knee, and forced the fractured end quite thro' the surrounding flesh, into the frozen ground. He expired in about two hours, leaving a wife and eight children to lament his loss."[7]

If falling trees and branches didn't injure a chopper, by the 1950s a chain saw likely would eventually, the average chain saw–related injury requiring over one hundred stitches. While chain saw injuries to the lower extremities have been most common, those to the upper torso, neck, and face can be devastating, if not fatal—a chain designed to cut wood at high speeds impacting the less durable human body. The post–World War II transition from hand saws to gasoline-powered chain saws did not appear to some to require much training, despite the necessary skills and obvious dangers associated with chain saws. "Certainly, literacy was not required—so it was easy," wrote one Québécois observer of logging during the period. "The 'strong back and the weak mind' were adequate qualifications. The forest industry was not yet fully aware—at the end of World War II—of the absolute essentiality of well-organized training programs for Woods Workers," especially as it related to logging safety around the most dangerous hand tool of the period.[8] Unfortunately, although logger training programs, especially in the area of chain saw safety, are now common in the region, evidence suggests that such training alone may have little influence over chain saw injury rates.[9] This is likely due to some training programs' often too simplistic and narrow focus on safe technique, with little or no attention paid to the assessment of risk and the consequences of risky behavior, especially among loggers who have engaged in such behavior for years and have lived to boast about it.

DONE BY FORTY

Although episodic and traumatic, sometimes fatal logging injuries are more often reported, mythologized, and even romanticized, the chronic physical debilitations associated with the daily work functions of a hand-crew logger—felling and delimbing trees, pulling cable, setting chokers, operating woods equipment on difficult terrain—often go unnoticed. According to a Bangor, Maine, chiropractor interviewed in the late 1970s for the documentary film *Cut and Run*, "We see a considerable amount of woodsmen in our office . . . possibly 40 percent of the patients who come in through the office are woodsmen who have been hurt on the job . . . [They] are physically ready to retire when they get to be 35, 40, 50 years old."[10] A few years later, a publication of the Maine Department of Conservation titled "Maine Woods Labor Study, 1984–85," drew similar conclusions, suggesting that "many people who work with loggers comment that after 20 years in the woods, or about age 40, they are completely 'burnt out' . . . The low status of the occupation, the low pay, and the backbreaking work combine to favor other employment."[11] For those northern forest woods workers who lived in logging camps, there were reports of health issues related to diet, fire safety, and hygiene sprinkled with the inevitable drollery about bunkhouse life. "Il n'ya pas de poux—les punaises les ont tout mangé [There are no lice—the bedbugs have eaten them all]," went one witticism.[12]

The impact of injuries goes beyond the immediate victims, affecting families and the small rural communities in which they live. The effects can be devastating, especially considering that few employees in the northern forest have work-related health benefits beyond what is required through workers' compensation insurance. In addition, the separation from one's family and the anticipation of logging injuries, along with the method of pay used to compensate some loggers, can take a toll. The 1980 documentary *Cut and Run* interviewed a Maine logger who appeared to lament the transition from time-based to piece-rate wages, a system that, according to some, was a common cost-cutting shift among Maine's paper mills in the late 1920s and 1930s: "I think I move a lot faster than the guys did in the old days because what we get for our wood is worth a hell of a lot less . . . in order for me to get a good day's pay, I gotta cut 120 trees alone, and in the old days, well, they didn't have to cut that many trees." Echoing this logger's anxiety about the inherent dangers of logging, Lucien LaChance, a Québécois logger working for Scott Paper Company in Maine, talked about his wife's concern for his safety: "You know my wife has been telling me for at least four or five years: 'Lucien, when are you going to stop going in the woods? I'm always afraid of the phone call that you've had an accident at work from a stub or a tree and that you're hurt or dead. When are you going to stop working in the woods? I long for the day when you'll be able to have a life with us here together.' I always tell her I'll stop, but I never do." Reflecting on the difficult life of a northern forest logger trying to make a living earning piece-rate wages, LaChance continued: "We work at a speed just to make a decent wage, and at a certain age there will come a time, whether you are Canadian or American, when you won't be able to work this hard, to keep up this pace—it's impossible." In a plea for a more unified logging community, one that recognizes the stresses associated with a system of piece-rate wages prevalent among pulpwood logging operations in northern Maine at the time, he continued, "I'd say 'Guys, take the time to work safely. It's an answer to the fast pace that our lives are going in. Let's join hands. Let's slow down the pace so we can work longer together. In ten years, at this pace we're working at now, we won't be able to continue.'"[13]

But "joining hands," especially across borders, has not necessarily been a strength of the region's logging community. A northern New Hampshire logger asserted during a group meeting of loggers from the Granite State in 2000, "We're our own worst enemies because we've overproduced." Casting doubt on whether the logging community of the northern forest could ever truly work together in such an isolated, competitive, high-risk industry was a Maine logger, who during a group meeting held in Bangor the same year claimed, "As long as you have independent loggers, we won't stick together." Indeed, as we shall see later, the "independent logger" phenomenon would become a source of pervasive tension in the state's logging woods.[14]

TREE, SPARE THAT WOODSMAN!

The following edited and annotated deposition, taken from a logger injured during an accident in the northern forest in the 1980s, provides insight into the dangers of logging as well as logging practices and culture of the period, most of which have persisted into the twenty-first century. The injured logger, a plaintiff seeking damages related to his injury, is being questioned by an attorney representing the defendant, a company responsible for managing the forest ownership being harvested. Poorly educated and chronically down on his luck, he boasted that he had logged all his life, an assertion not atypical of the hyperbole and bluster expressed by many whose livelihoods depend on the next logging chance, the next woodlot to cut. He likely knew nothing and cared little about logging lore—of bateaus, logging camps, log drives, and misery whips.[15] He was just trying to make a living, feed himself and his family, doing the most dangerous work around. Now wheelchair bound, he had been a chopper—making a living too close to the stump—and had paid the price. Pivoting on a single bad decision, it all went haywire that day, the chopper's already hard life changing forever. Although unclear from his deposition, he may have made the common mistake of attempting to fell a yellow birch tree that was supporting a red maple he had tried unsuccessfully to bring to the ground with his chain saw.

Whether he tripped and fell or just never really had a chance, nobody knew, he didn't remember, and the logging woods were indifferent. His broken body was pinned under what a split second before was an unexpected cacophony of falling, crashing trees and dislodged branches. Leaves, upset by the commotion, quietly cascaded around him and what he had done, blanketing him with the red, orange, and yellow hues of the damp and now suddenly silent northern forest in autumn. The injured logger sued both the landowner (a forest management company) and the logging contractor for compensation, the plaintiff's lawyer asserting that his client was injured by a hazardous tree. The landowner, he claimed, was responsible for this preexisting defective condition on the property.

The words of the injured man during his deposition offer some insight into the day-to-day reality of life in the logging woods. The words of the injured logger and the attorney are preserved below, but not all questions and answers are recorded here, some extraneous narrative is deleted, the actual name of the logging company has been changed to White Pine Logging, and the injured logger has not been identified. Otherwise, questions and responses are verbatim from transcripts of the deposition.

The deposition begins with questioning of the injured logger by the attorney for the defense, followed by the injured logger's responses:

DEFENSE ATTORNEY: I'm going to ask you some questions about an accident which occurred . . . and your background, and the injuries that you sustained in this accident. If for some reason you don't understand a question that I ask you, would you please let me know so I can clarify.

PLAINTIFF: Yes.

DEFENSE ATTORNEY: And keep all of your answers oral; speak them, don't shake your head.

PLAINTIFF: Yes.

DEFENSE ATTORNEY: Where were you employed at the time of the accident?

PLAINTIFF: I was employed by White Pine Logging.

At the time of the accident, almost all logging employees in the northern forest region worked for independent logging contractors, not large, industrial forest management companies or wood- and fiber-consuming mills. These logging contractors typically entered into contracts with large landowners to provide logging services. This employment was a change from earlier times when many loggers in the region were employed directly by large forest products companies, especially in northern Maine and proximate to paper mills throughout the northern forest.

However, there have also been independent loggers throughout the region who have never been employed directly by paper companies and who worked directly with forestland owners, typically buying stumpage from them.[16] This is a common timber transaction model in the more southern part of the region where nonindustrial private forest owners (versus industrial owners, such as paper companies) own most of the timberland and generally engage in stumpage sales rather than in contracts for logging services. The process of selling stumpage could involve a timber cruise (an inventory of the trees to be harvested); a prospectus (a document that summarizes the results of the cruise, location and attributes of the logging chance, expectations of the seller, such as utilization standards and road building, and any restrictions, for example, use of certain logging equipment); a showing, during which an agent of the landowner shows the logging chance to prospective buyers; and a sealed-bid auction. The successful bidder enters into a contract with the seller, the contract spelling out terms of the sale. Alternatively, this may all be accomplished with a simple handshake between buyer and seller. The actual sale of stumpage typically takes one of two forms: lump sum, in which a price for standing timber is paid to the landowner up front, before logging commences, based on an estimate of the volume and value of the timber to be harvested; and unit price, or "pay as you cut" sales, in which the landowner is paid as the timber is cut, scaled, delivered, and paid for by a consuming mill or log broker.

Log scaling in the northern forest is usually performed at the point of sale by a log scaler, usually on board-foot basis. Some loggers may also scale a load of logs before it leaves the log landing, especially if they have reason to believe that the scale at the point of sale has been inaccurate. The actual log rule used is often peculiar to a region or mill. Although dozens of log rules exist, the International ¼-Inch Rule is considered the official log rule in Maine, but other log rules can be used if the buyer and seller agree. The Scribner Rule, which underestimates the International ¼-Inch Rule, prevailed where this injured logger was chopping. Logs may also be bought and sold based on grade: their quality relative to their use for higher- (e.g., veneer and sawlogs) or lower-value (e.g., tie, pallet, and round edge logs) wood products. Biomass, generally in the form of wood chips, and some pulp logs are bought and sold based on weight, although the cord (128 cubic feet of stacked wood), used to measure firewood, may be used for pulp. The so-called face cord (dimensions vary, but with a stacked volume less than 128 cubic feet) was more common where this accident occurred.

DEFENSE ATTORNEY: What was your position for White Pine Logging?
PLAINTIFF: I was a faller.

In northern New England the word *chopper* is often used to describe the person who performs the same work functions as a *faller*—primarily tree felling and delimbing. One may also hear the words *feller*, *sawhand*, and *sawyer*, the latter more common in the southern United States.

DEFENSE ATTORNEY: How long had you been employed with White Pine Logging?
PLAINTIFF: About seven years.
DEFENSE ATTORNEY: For how many years have you been generally engaged in logging work?

PLAINTIFF: Well, I started in the business when I was 18 years old. That's the only trade I really know.

DEFENSE ATTORNEY: So, you would have been working in the logging industry for approximately 25 years before you had your accident. Is that a fair statement?

PLAINTIFF: Yes.

Research from the University of Maine showed that the average age of loggers in the northern forest region in 2004 was about forty-six years, while the average number of years logging was about twenty-three, although there is a great deal of variability in these attributes across the northern forest region. For example, in the Adirondacks in 2006, the average age of logging business owners was fifty, and the average number of years logging was twenty-eight.[17]

DEFENSE ATTORNEY: What exactly does a faller do in the logging business?

PLAINTIFF: A faller goes in and he looks around. He makes an escape route before he falls a tree. Then he falls it and limbs it.

There are fairly standard procedures that choppers follow for the safe and efficient felling of trees, including assessing the tree for hazards; evaluating the tree for lean; planning an "escape" from the tree before it falls; and determining the best "lay" for the tree so as to avoid, for example, forest regeneration and residual trees, and to facilitate efficient skidding. These determinations are critical to safe and efficient chopping, and not one is necessarily easy to make or foolproof. For example, a tree that appears to be free of hazards may indeed have a loose branch, or "widow-maker," in its crown that cannot be easily detected from the ground by a chopper.

The Occupational Safety and Health Administration (OSHA) promulgated a comprehensive standard for logging for the first time in 1994, despite the dangers that had long been associated with logging and logging work functions. Prior to that, a 1971 OSHA pulpwood standard, the sometimes elusive "standard industry practice," and standards established by the American Pulpwood Association guided much of logging safety in the United States and the litigation that inevitably followed when a logging accident occurred.

DEFENSE ATTORNEY:	Was that your general job description, the thing that you did for White Pine Logging?
PLAINTIFF:	Yes.
DEFENSE ATTORNEY:	And how much did White Pine Logging pay you for a day's work?
PLAINTIFF:	Offhand, I wouldn't figure it out. I would have to look back.

It is not at all uncommon for a logging employee to be paid based on time, although some are paid based on production. However, many believe that transitions from daily or hourly wages to piece-rate wages came at a price. According to one observer, "nearly everyone, including insurance companies, agrees that piece-rate pay is a major culprit [in compromising logger safety]. It pressures loggers to speed the pace of work, to cut when they are fatigued, to work in treacherous conditions and to neglect cumbersome safety apparel."[18] According to the U.S. Census, the median income for full-time loggers in 1980 was about $11,000, and during the same year "over one-fifth of loggers' households (twice the national average) lived in poverty."[19]

DEFENSE ATTORNEY:	What sort of things did White Pine Logging take out of your paycheck for?
PLAINTIFF:	Took out for Social Security; took out for state and federal tax. Gee, I don't—I really didn't pay much attention what was being took out.
DEFENSE ATTORNEY:	Did he give you medical insurance?
PLAINTIFF:	Just [workers'] compensation. That's all the medical we had.
DEFENSE ATTORNEY:	Did he give you any other benefits aside from the money that he gave you in your paycheck?
PLAINTIFF:	Just a paycheck.

Although this is changing, especially among the region's better-organized and more highly evolved logging companies, typical employment benefits were not common among logging employees in the northern forest region in the early twenty-first century. Most logging employees in the northern forest region do not receive employment benefits, such as paid vacation and health coverage, beyond what may be covered by workers' compensation insurance. For example, a study of the logging workforce in northern New England in 2004 reported that only 16 percent of logging employees in the region received health insurance benefits and 21 percent received vacation time. Loggers from Vermont were the least likely to report receiving employment benefits (only 3 percent), and loggers from Maine were the most likely to report them (18 percent).[20] That employment benefits in the logging industry are poor or lacking is a major hurdle to the recruitment of new loggers. A 2009 survey of the general public in the state of Maine, for example, found that logging was not appealing as a form of employment to almost three-quarters of respondents because of their perception that the profession offered poor job benefits.[21]

DEFENSE ATTORNEY: And, approximately, how many days a week did you work . . . before your accident?
PLAINTIFF: Sometimes, it was five; sometimes, it was six days a week.
DEFENSE ATTORNEY: Did that have to do with the weather?
PLAINTIFF: Yes.
DEFENSE ATTORNEY: And if there was a bad day, really heavy rain or thunderstorms or something, you wouldn't work?
PLAINTIFF: We wouldn't.
DEFENSE ATTORNEY: And otherwise, you worked on a steady basis?
PLAINTIFF: Yep.

There is no typical, region-wide logging workday or workweek. In northern New England and elsewhere, markets, availability of stumpage, weather, and ground conditions—especially during mud season when the ground is soft from spring thawing—generally define the length and the rhythm of the logging year. For some northern forest loggers, mud season can last weeks—even months—depending on the onset of spring breakup and the amount of snowmelt and spring precipitation; for others, moving logging equipment to cut timber growing on sandy eskers, as in the Ossipee, New Hampshire, region, can help maintain logging operations through much or all of this period, as it did for the crew that this author chopped for in the 1980s. Mechanization, especially cut-to-length logging systems, has helped some loggers extend the logging season. But such systems are expensive and out of reach for many loggers in the northern forest. When the New Hampshire logging "hand crew" for which I worked in the 1980s comprised two cable skidder operators and a chopper, it worked Monday to Friday from 7:00 a.m. to 3:30 p.m. When the crew transitioned to a more mechanized biomass logging outfit, adding two feller-bunchers, two cable skidders, and a whole-tree chipper, workdays were often longer and the crew worked more days per week and more weeks each year. The crew was also paid better.

Despite the attention to logging mechanization in the northern forest, studies published in 2004 and eight years later in 2012 confirmed that most logging businesses in the region were conventional or hand crews, with felling by chain saws and in-woods transport by cable skidders.[22] Because they are relatively labor intensive, these systems, where trees are felled by hand, delimbed and topped, and then dragged to a landing to be bucked into logs by either chain saw or a more mechanized slasher, account for a relatively small proportion of the region's total wood production. However, there were significant differences in mechanization among the states comprising the northern forest, with the industrial pulpwood forests of northern Maine far more mechanized than elsewhere in the region. Most logging businesses in the region remained relatively small, the 2012 study finding that 86 percent of logging businesses in Maine, New Hampshire, New York, and Vermont had five or fewer employees.[23] Highlighting some fundamental differences between logging businesses in southern and northern Maine, a 2001 University of Maine showed that, on average, northern Maine logging businesses were more highly capitalized and worked more weeks per year than their southern Maine counterparts, both attributes suggesting more logging mechanization in northern Maine (see table 1).[24]

TABLE 1. Attributes of logging businesses from southern and northern Maine counties in 2001

Attribute	All Maine loggers	Southern Maine	Northern Maine
Number of weeks logging per year	39	36	40
Number of hours worked per week	57	44	60
Distance traveled to work (miles)	31	21	34
Capital invested in logging ($US)	299,514	156,424	346,089

SOURCE: Deryth Taggart and Andrew Egan, "Occupational Choice among Loggers in Maine's Northern Forest and Southern Counties," *Adirondack Journal of Environmental Studies* 9, no. 2 (Fall/Winter 2002): 33.

DEFENSE ATTORNEY: Were you involved in an accident?
PLAINTIFF: Just that one.
DEFENSE ATTORNEY: Okay. Where did that occur?
PLAINTIFF: It was on the—where I was cutting.

From 1980 to 1989, the period during which this 1985 accident occurred, there were 1,492 logging-related deaths, a fatality rate of 164 deaths per 100,000 workers—a rate that was 23 times that for all U.S. workers. Most of those deaths were associated with felling trees.[25]

Perhaps not surprisingly, data from a 1984 study of injuries in the logging industry inform us that the most dangerous workplace on a logging operation is the tree felling site. In general, the closer one worked to the stump, the more hazardous the work, although accidents can and do occur throughout the area being logged, often from being struck by or caught in between machinery, such as skidders and log loaders. Information from the Bureau of Labor Statistics around the time of this accident showed that 53 percent of all logging injuries occurred at the stump (tree felling, delimbing, topping), and 18 percent took place between the stump and the landing. In addition, during the time of this accident, data show that the act of tree felling was responsible for nearly one-quarter of all logging injuries—most often from falling trees.[26]

DEFENSE ATTORNEY:	When you were working . . . were there people that worked with you for White Pine Logging?
PLAINTIFF:	Yes, there was; skidder driver and header man.

Referring to the relationship between choppers and skidder operators as elements of a "specialized assembly line process," one observer of logging in the northern forest asserted that the skidder was "disruptive to traditional patterns of work organization," because many pulpwood cutters had essentially been independent workers who were paid based on what they felled. But "mechanized skidding produced a division of labor. Rather than having fellers buck the trees at the stump, the bucker performed this task at the landing," as skidders allowed for tree-length dragging to the landing. "In essence, the skidder set the pace for a number of cutters. Although cutters still organized their own work, they now had to gear their rate of production to that of the skidders, which traveled faster than horses and required no rest. Pressured to keep up with skidders, fellers and buckers had less control over their work, which was a source of some discontent."[27]

Rubber-tired cable skidders are still common in the northern forest, despite advances in mechanized felling, grapple skidding, and forwarding. A survey of logging businesses in northern New England in 2006 showed that most skidders operating in that region were cable skidders, which require the operator to pay out and pull cable by hand from the back of the skidder and set wire or chain chokers to felled wood for subsequent dragging to a landing. It's common for hand crews to comprise a faller with a chain saw and one or more cable skidders. Grapple skidders are often, and more efficiently, teamed with mechanized felling, such as feller-bunchers, a system that has become more and more common in the region given its higher production rates and often lower workers' compensation insurance costs than more manual logging methods.

This deposition refers to the operator of a rubber-tired cable skidder, and the logging crew would be considered a conventional, or "hand," crew as opposed to a mechanized, or "mechanical," crew where trees are cut by a feller-buncher and dragged by a grapple skidder, for example.

DEFENSE ATTORNEY:	What does a skidder driver do?
PLAINTIFF:	Well, a skidder driver does—he goes in and hooks up the trees and takes them out to the landing.

FIGURE 4. In the post–World War II logging woods, chain saws replaced axes and crosscut saws; cable skidders (left) and, later, grapple skidders (right) replaced horsepower. The high cost of logging mechanization turned newly independent logging contractors into small and medium-sized business owners who competed for work in Maine's northern forest. Photograph by the author.

DEFENSE ATTORNEY:	Does that with some kind of machine?
PLAINTIFF:	Yes; with a skidder.
DEFENSE ATTORNEY:	Is that something you ride on?
PLAINTIFF:	Yes.
DEFENSE ATTORNEY:	Okay. And you pull the logs with chains?
PLAINTIFF:	Well, it has got a cable, and you wrap the chains around the logs.

Cable skidders, essentially heavy-duty woods tractors that are purpose-built for the rough forest conditions associated with logging, have a cable wound around a drum and to which are attached slides, or metal collars, that can run freely up and down the cable carrying hardware to which cable or chain chokers are attached. Chokers are used to attach (or "hook") to felled, delimbed trees, which are then skidded to the landing for cutting into logs. The process of hooking a choker to a felled tree or log is often referred to as *choker setting.* Sometimes tree-length wood is cut into logs in the woods if smaller sections must be skidded, for example, when a small farm tractor or horse is used to skid wood and tree-length wood is too heavy to drag or when tree-length wood is too long for a tight network of skid trails being used to transport it from the stump to a staging area or log landing.[28]

DEFENSE ATTORNEY: Okay. What does a header man do?
PLAINTIFF: Saws up the trees after they are brought out.

A header man works on the header (aka, landing, deck, brow), a cleared staging area where felled wood is bucked into logs and loaded onto trucks for transport to a mill or concentration yard), typically cutting ("bucking") tree-length wood into logs, with log lengths depending on market specifications. For example, if wood markets demand sixteen-foot construction material, logs would be bucked to sixteen feet, with a couple of inches of trim over that length to account for damage to the ends of the logs from skidding, decking, and loading. At one time, white birch in one region of the northern forest was bucked to nine- or thirteen-foot logs, such that once delivered to the mill, the logs would be recut into four-foot sections—"bolts"—to be turned for veneer, the extra foot comprising log trim. Softwood pulp is still sometimes cut to four-foot lengths—a throwback to the ease of transporting four-foot logs on rivers, with manufacturing processes at the receiving mills downriver built to accommodate this length. More modern pulp mills will take tree-length pulp logs, which are more efficient to manufacture (i.e., there are fewer cuts on the landing to make tree-length wood into logs), load at the landing, and unload at the mill.

Bucking tree-length wood into logs on a landing is done with a chain saw or with a more mechanized slasher. However, some mechanized in-woods systems fell, delimb, and buck using an in-woods processor, sometimes referred to as a *harvester*, typically followed by a forwarder to perform in-woods transport to a landing.

DEFENSE ATTORNEY: Now, in September of 1986 in the area where you were cutting wood, what kind of wood were you cutting, hardwood?
PLAINTIFF: Hardwood.

In this area and throughout much of the northern forest, hardwood (or deciduous) species were typically sugar maple, red maple, yellow birch, white birch, and American beech. Softwood (or coniferous) trees found in the region include eastern white pine, eastern hemlock, balsam fir, and red and black spruce.

DEFENSE ATTORNEY: Was it pulpwood? Was it for furniture, things like that?

PLAINTIFF: It was pulp. And the good trees we saved logs out of the butts of them.

In the northern forest it's not unusual for a single logging chance to yield pulpwood (generally smaller, lower-quality trees and logs), sawlogs, and veneer logs and for the chopper to be felling pulpwood trees one minute and sawlog or veneer log trees the next, depending on markets and the stand being logged. The same tree may contain pulpwood logs—especially toward the top of the tree where log diameters are smaller and logs are of lower quality—sawlogs and, typically at the butt (big end of the tree), veneer logs. Depending on log quality, so-called butt logs are the largest and often most valuable logs that can be cut from a tree.

If asked what he is cutting, a chopper might respond by the general type of tree (hardwood or softwood) or by the species being harvested (e.g., soft maple, pine, etc.). Trees and logs are also often referred to by their intended use. For example, pallet logs were typically low-quality hardwood logs that were eventually processed into pallets. Softwood and hardwood trees that have the size and quality to be cut into sawlogs, or *sawtimber*, are then sawed at a mill into boards. The jargon may shift from one locality to another in the northern forest, depending on local logging culture and markets. For example, there may be a local market for live-sawn lumber, that is, boards that are not square-edged but sold with round edges. This is true of some white pine in the region, where round edge is sometimes used for siding, for example. Logs used for round edge are generally smaller, lower-quality pine sawlogs, and these and the trees from which they are cut may also be referred to as "round edge" because that is their intended use. Other trees may be used to produce "shook," smaller pieces of lumber, rather than entire log-length boards, that at one time were used to manufacture wooden cable reels, for instance.

DEFENSE ATTORNEY: How would you know which trees you should cut down?

PLAINTIFF: Well, the trees that had the paint on them were the ones we were supposed to cut.

A forester will typically mark with paint the population of trees—those to be cut versus those to be left—that is the smaller. Some operations, especially for the removal of trees for biomass, leave fewer trees than are harvested; for efficiency, the leave population of trees is marked. In the case of this accident, the faller was cutting marked trees.

However, a conundrum related to marking, logger safety, and liability plays out in this logging accident and many others. The attorney posing questions during this deposition is likely asking whether the trees were marked to determine the degree to which the forester was giving direction to the logging crew through his paint marks. If it can be demonstrated that the forester was providing sufficient direction to the chopper, the case could be made that the chopper was behaving as if he were an employee of the land management company for which the forester worked and that the land management company was therefore liable for the accident. I will discuss these distinctions, a source of significant conflict in Maine's logging woods, in the next chapter.

Foresters will often mark dangerous situations in the woods as well, such as trees with widow-makers (loose branches and treetops and leaning or hung-up trees that could fall and injure a chopper), to warn the chopper to look up. However, if the timber marker does not notice a widow-maker that later injures a chopper, a suit against the timber marker or, more likely, the forest management company he represents or even the landowner, is sometimes brought. For example, in 1992, a timber faller claimed to be struck by a snag that was accidentally tipped in his direction when the skidder operator was pulling down a severed tree that the faller had lodged in a nearby supporting tree. The faller, an employee of an independent logging contractor, alleged that he was injured by the snag, which his attorney claimed was a preexisting dangerous condition that the landowner should have noted and neutralized before the logging operation commenced. The suit was settled out of court.[29]

Fear of liability can impact forestry field practices and, unfortunately, compromise the important sense of teamwork among landowner, forester, and logger during harvesting. One woodlands manager from a large Maine paper company stated, "We are discouraged by our attorneys from doing too much in the way of safety, for instance, because of the danger of destroying the independent relationship. If you control their operation you diminish their independence, and the federal government says that they are your employees."[30] That is, the wood-consuming mill may become liable for injuries to loggers whom they direct on the job, whether the mill actually employed the injured logger or not.

According to one study of logging-related accidents, "direct communications with woods workers, especially communications that describe how trees should be cut or skidded, may create a perception that the landowner or forester, rather than the contractor, is in the position of employer . . . perhaps holding the forester or landowner liable for injuries that may occur to the woods worker . . . [W]ill the forestry community respond to expensive, time-consuming litigation by distancing themselves from loggers to emphasize that an employer-employee relationship does not exist between them, as was recently suggested by an industry forester? Will foresters now refrain from indicating logging hazards they find during the planning or marking of timber harvests so that they may be protected in the event they overlook a hazard that later injures a logger? One consulting forester, who used to indicate hazards with 'killer tree' flagging, no longer does so for this reason."[31]

Unfortunately, hazard trees comprise at least twenty trees per acre in the central Appalachian region that includes the northern forest of the state of New York, making it virtually impossible to account for all hazardous trees and tree conditions, especially those related to compromised root systems. Among the tree species most commonly exhibiting dangerous conditions, such as broken limbs and tops, are yellow birch, red spruce, white pine, and American beech—familiar tree species in Maine and the rest of the northern forest region.[32]

Nevertheless, there are those who believe that professional foresters should be considered employers of independent logging contractors, essentially redefining the forester–independent logging contractor relationship. For example, the authors of a *Journal of Forestry* article on the implications of the OSHA logging standard to forest managers asserted that "logging operators are clearly constituents of forest managers" and that "the forest manager has a professional obligation . . . to assist the logging operator in making any operation as safe as possible." They suggested that "in many instances, the forest manager accepts the role of primary contractor when administering a timber sale for a landowner" and that "forest managers should be aware that when they accept the role of primary contractor, they also *potentially* accept some level of legal responsibility for the identification and control of the logging hazards defined in the OSHA standard." The authors, employees of the National Institute for Occupational Safety and Health, were not specific about in which "instances" this responsibility should apply nor what the scope of a forester's presumed "level of legal responsibility" might be.[33]

Whether independent logging contractors are truly independent or are operationally behaving as if employees of larger industrial forestland management companies and wood-consuming mills is a basis of much of the tension between loggers and the forestry sector in the northern forest and elsewhere, as we shall see later. The type of timber sale may also be important in understanding liability. When the sale is lump sum, the stumpage buyer—in this case, the logging contractor—owns the standing timber before it is cut and therefore, it may be argued, owns the risk associated with harvesting that timber. However, for unit price or pay-as-you-cut stumpage sales, the landowner owns the timber until it is severed and may be shown to be responsible for any "defective condition" in the stand.[34]

Adding to the complexity, a 1999 University of Maine study concluded that most nonindustrial forest owners sell stumpage as trees are harvested and delivered to a wood-consuming mill, most often doing so without the assistance of a forester or a written contract between the landowner and the logging contractor.[35] Another study found that less than half of all partial harvests that were administered by foresters in the state of Maine were marked prior to cutting.[36]

DEFENSE ATTORNEY: Did you ever have any kind of specialized training? Did you ever go to school or have any training through your employer on how you should do your job?

PLAINTIFF: No. It was all—

DEFENSE ATTORNEY: It was all on the job?

PLAINTIFF: Yes.

DEFENSE ATTORNEY: Okay. And you had experience with this particular job with falling trees; isn't that right?

PLAINTIFF: Yes.

DEFENSE ATTORNEY: And you used a saw when you did that?

PLAINTIFF: Yes.

The chain on a production chain saw rotates around a guide bar at about sixty miles per hour, and the cutting links on these chains are designed and filed to sever wood fibers at these high speeds. Although professional chain saws are equipped with antikickback mechanisms, these devices merely arrest some of the energy associated with chainsaw kickback by stopping the chain from spinning around the bar. They don't really stop kickback—the back-rotation of a saw around a stalled saw chain rather than a chain rotating forward around the guide bar. This back-rotation can be in the direction of the operator, depending on how the saw is being used. Antikickback devices also don't always engage with every kickback, depending on how well they are maintained and adjusted and how the saw is being handled by the operator. Every logger who uses a chain saw experiences chain saw kickback at some time; luckily, few are injured. When a kickback injury does occur, the results are often devastating, especially if the energy of the kickback is in the direction of the operator's face or neck. A chain saw engine also whines at over one hundred decibels, and permanent hearing loss can occur after less than two minutes of unprotected exposure.

DEFENSE ATTORNEY: In the time that you worked as a faller, did you ever come to recognize or has anybody ever pointed out to you certain hazardous conditions that might exist in trees in doing that type of work?

PLAINTIFF: No. You learn when you go along. You pick it up.

DEFENSE ATTORNEY: So, you learn that certain things are dangerous; isn't that right?

PLAINTIFF: Yes.

The year before this accident occurred, the Bureau of Labor Statistics reported that over half of all logging-related injuries took place at the felling site and 18 percent between the stump and the log landing.[37] A decade later, researchers found that most logging industry–related deaths occurred to workers involved in the felling, preparation and skidding of logs and trees.[38] Comparing logging-relate deaths to those that occur across the manufacturing sector, the following year the same authors cited "in-house" Department of Health and Human Services data that showed that, although logging accounted for only about 0.4 percent of the manufacturing employment in the United States during the period of this accident (1980–88), it accounted for approximately 20 percent of all deaths across all U.S. manufacturing sectors—an average of 161 deaths per 100,000 full-time workers. The total private sector rate was only seven deaths per 100,000 full-time workers during approximately the same period.[39]

Today, there are many logger training and certification programs in the northern forest, but none were active at the time of this accident, and none of the region's four states required logger training. The state of Maine has perhaps the most mature and comprehensive training program: the Certified Logging Professionals Program (CLP), founded in 1991 and headquartered in Jackman, near the Québec border. Those companies that train their employees through CLP's four-day program can have their workers' compensation rates reduced. However, while injuries and illnesses per one hundred full-time workers have decreased since the beginning of CLP, it is sometimes difficult to separate the precise role of the program versus other factors—for example, increases in logging mechanization—that may have reduced the exposure of employees to hazards.

Logger certification training can be expensive, but so are logging-related accidents. The average cost of certification per logging crew member in West Virginia in 1997 was $1,617.69,[40] with loggers' highest level of dissatisfaction being chain saw safety training (versus training in best management practices and first aid). The most experienced loggers were found to be least satisfied with that state's certification program. Negative feelings about the cost of logger training among logging business owners was summed up by one West Virginia logger who asserted that the cost of training was "too damn much!" In addition to the costs associated with safety training, who designs and delivers such training matters to some loggers. "The forestry community has no conception of the safety aspects of logging," asserted one Connecticut logger during a group meeting in 2005.[41]

It is undeniable, however, that workers' compensation rates for logging businesses in Maine have fallen, from $45 per hundred dollars of payroll in 1993 to approximately $20 per hundred dollars of payroll in 2013 for "conventional crews," those crews that used chain saws to fell and delimb and skidders, often cable skidders, to drag wood from the stump to a landing. Although this was the same rate regardless of whether it was a CLP- or non-CLP-trained crew, there were large rate differentials between certified and noncertified crews through much of the late 1990s and early 2000s. In 2013, for example, the non-CLP rate was $13.24, whereas the CLP rate was only $4.13 per $100 of payroll.

All of these programs were voluntary. However, significant for these programs was that the Sustainable Forestry Initiative (SFI), an industry-sponsored forest certification program, required SFI member mills to purchase wood from loggers who had completed a nationally accredited logger education program. These programs met that criterion. None of these, however, existed at the time of the accident being discussed, and any safety training that may have existed was rather sporadic and poorly developed.

The injured logger in this deposition did not receive formal logger training. Even if he had, research from West Virginia suggests that participation in safety training does not necessarily reduce logging accidents.[42] Results suggested that logging employee turnover may be a more reliable variable associated with the efficacy of logger training, because it was found that workers' compensation insurance claim rates for loggers declined significantly the longer they worked for a company. And turnover in the logging workforce is complex, pivoting on more structural and systemic variables such as adequate pay, benefits, and a safe work environment, which we'll explore in chapter six.

In addition, most training focuses on technique, generally overlooking the motivation of a logger to take certain risks to personal safety to be more productive, potentially increasing their pay. It is common to hear logging safety instructors—often well-intentioned people, sometimes with little or no logging experience—say that "safety pays." Unfortunately, the experienced logging employee knows that taking risks may also pay in increased productivity, otherwise he would likely not do so. As expressed by one Swedish researcher writing in 1984 about logging accidents and their relationship to piece-rate wages, "It pays to take a risk. At least most of the time, since risk-taking or unsafe behavior is rarely followed by an accident. On the contrary it is often, or is experienced to be, the easier, less time-consuming, less strenuous, more comfortable and productivity increasing way of doing things," he suggested. "By what means can you convince a person, who has been applying 'unsafe' methods for twenty years without having an accident, to change to safer but less rewarding work habits? The only reward you can offer him is the probability that he will not be hurt in the future. This is no reward for him since he does not expect an accident anyhow from his unsafe behavior."[43] So the challenge is convincing loggers who have successfully and productively practiced unsafe behavior that their risk-taking somehow doesn't pay—not an easy proposition, since such behavior is seldom followed by accidents but is often rewarded with increased efficiency, even increased earnings in cases where pay is pegged to productivity. Although there are inherent safety risks to logging, the remaining challenge is around the best ways to modify unsafe behavior among loggers.

DEFENSE ATTORNEY: And what are some of the things that are dangerous in your line of work as a faller?

PLAINTIFF: Well, you look at a tree. You see if there is any dead limbs in it or if there's a dead top in it, or if there's anything cut in it. Then when you start cutting it, if the saw goes too easy, you figure it is a hollow-hearted tree and you be more careful.

In general, OSHA defines a "danger tree" as a "standing tree that presents a hazard to employees due to conditions such as, but not limited to, deterioration or physical damage to the root system, trunk, stem, or limbs, and the direction and lean of the tree." Such trees must be felled, removed, or avoided, according to OSHA. If avoided, no work should be conducted within two tree heights of the danger tree in question.[44]

A hollow tree or one with significant decay can be dangerous to fell because a chopper relies on sound wood to form a hinge to help guide the tree into a predetermined, prepared lay, avoiding residual trees (those trees left standing after the logging operation) and advanced forest regeneration, for example. Without sound wood along this guiding hinge, the tree's fall is much less predictable, making it more dangerous to cut. Whether any of these conditions and behaviors were in play in this accident is unknown.

DEFENSE ATTORNEY:	Okay. So, dead tops and dead limbs and hollow trees, those are three things that you look for.
PLAINTIFF:	Uh-hum.
DEFENSE ATTORNEY:	Do you look for hazards around you?
PLAINTIFF:	Yes. You look for hazards around you.
DEFENSE ATTORNEY:	Like what?
PLAINTIFF:	Like trees laying into them, and stuff.

Such trees are often referred to as "hang-ups" or hung trees—extremely dangerous in the logging woods.

DEFENSE ATTORNEY:	So, other trees that have fallen against the tree that you're cutting?
PLAINTIFF:	Yes.

Unfortunately, so-called danger or hazard trees are common in the northeast United States. A broad forest inventory conducted in 1998 showed approximately twenty-one danger trees per acre in the forests of New York State, not counting those trees that might be considered dangerous because of unstable or compromised root systems.[45] Two-thirds of the danger trees were considered snags: standing dead trees. The most hazardous trees in the region are those in the northern hardwood timber type—birch, beech, and maple—the same type of trees that this faller was cutting when he was injured and a common timber type in the northern forest region.

DEFENSE ATTORNEY:	And do you look around for any of the other things that you have described, the dead tops and–
PLAINTIFF:	Always look around the tree. You always look behind you to see which is the best way out.
DEFENSE ATTORNEY:	And you always do that?

PLAINTIFF: Yes. Always.

DEFENSE ATTORNEY: Now, you mentioned earlier about an escape route. What do you do specifically to make sure that you have an escape route?

PLAINTIFF: Escape route—when you go up and look at a tree, you see which way it is going to fall. Then, you look on both sides of you and you look behind you to see which way is the best way to get away from that tree.

Before notching and backcutting a tree, safe felling practice involves planning an escape route, generally at an angle away from the intended direction of the falling tree. This path is typically cleared of anything that might get in the way of or trip the chopper (e.g., logging slash, small trees) during his retreat. The felled tree is not supposed to be approached for delimbing until the chopper is sure that the tree has been safely grounded and that there are no newly created widow-makers in neighboring trees.

DEFENSE ATTORNEY: And you recognize that logging is a dangerous business. Is that a fair statement?

PLAINTIFF: Yes.

As we've discussed, manual felling is the most dangerous of all logging-related activities, and falling objects account for most logging fatalities, most of which occur to choppers and skidder operators. Poor felling technique has been implicated directly in about 15 percent of felling-related fatalities, although when combined with poor decision making and risk-taking behavior on the part of the chopper, this figure is likely an underestimate.

DEFENSE ATTORNEY: Now when you do logging—when you do falling, are there any specific practices that you follow in terms of the way that you cut trees?

PLAINTIFF: Yes.

DEFENSE ATTORNEY: Okay. Do you make certain that the trees that you're cutting are going to fall?

PLAINTIFF: Yes.

A hinge of sound wood on the stump over which a tree pivots helps guide it in the desired direction—its "lay"—as it falls. Wedges are often used when felling, especially in situations where a tree must be lifted in the desired direction and away from its more natural lean.

DEFENSE ATTORNEY:	Do you make sure that they are not going to fall into other trees?
PLAINTIFF:	Well, if they fall on other trees, you leave them, and then you have the skidder get them out.
DEFENSE ATTORNEY:	How do you make sure that they are not going to fall into other trees?
PLAINTIFF:	Because you hang one corner or the other, and you try to pull it away from the other trees.

Loggers generally adopt the regional jargon around logging when describing their work. But some loggers and crews also have their own individual jargon. This injured logger is likely referring to the Dutchman cut, which is a way to manipulate the hinge of a tree to essentially change the lean of the tree, guiding it into a more desirable lay on the ground, one that might avoid regeneration or residual trees, for example.

DEFENSE ATTORNEY:	What happens when one tree falls into another one?
PLAINTIFF:	They hang up.

This isn't always the case. Some loggers will plan to fell a group of trees that are notched and back cut ahead of a pusher tree. These trees sit back on the stump, needing only a push from an "uphill" tree for them to begin to tip forward in the desired direction. This is called *domino felling*, which is frowned upon by OSHA and most logging safety programs.

Domino felling, although dangerous, is a form of intentional, choreographed directional felling that can be efficient, but dangerous. This injured logger describes a situation where a tree is felled forward, engaging another, uncut, tree that then hangs up and supports the falling tree, preventing it from falling to the ground. The unintended hanging up of a felled tree is accidental and inefficient, since it ties up the skidder to pull the hung tree to the ground, instead of dragging trees to the landing for bucking. Fallers are taught—or soon learn—never to work on a tree that is supporting a felled, hung-up tree. Many, however, still do, despite the obvious risks involved.

In northern hardwoods, as in the case of this accident, hang-ups can be more common in the winter when the branches are rigid and don't flex as they fall, as they do in the spring. As a result, branches of falling trees are more apt to be tangled in the branches of neighboring standing trees.

DEFENSE ATTORNEY: And when you hang up a tree, what is your experience in terms of the best way to handle that situation?

PLAINTIFF: You leave it. Then, when the skidder comes in, you hook on it and pull it down.

DEFENSE ATTORNEY: Have you, in the time that you have worked as a faller, ever handled a hung-up tree by cutting down the tree that the tree was hung up in?

PLAINTIFF: Well, not really.

DEFENSE ATTORNEY: And, as I understand it, you were paid by-the-week wages, right?

PLAINTIFF: Right.

DEFENSE ATTORNEY: Okay. Didn't make any difference if you cut one tree or twenty trees.

PLAINTIFF: Right.

DEFENSE ATTORNEY: Do some guys who work in the woods get paid by the board foot?

PLAINTIFF: Some jobs do it that way, yes.

DEFENSE ATTORNEY: But, in this particular job you were paid by the hour, sir.

PLAINTIFF: Right.

Some loggers are paid piece-rate wages, that is, wages that are based on production (per cord or per thousand board feet of logs skidded to the landing or delivered to a mill, for example), rather than on time worked—by the hour, day, or week. Piece-rate wages, sometimes referred to as "piecework" or "incentive payment," offered some advantages over hourly, daily, weekly, or monthly wages to those employing loggers, including less need for close supervision and encouraging "a competitiveness among the men that undermined solidarity. Workers themselves seemed to prefer remuneration on a basis that promised the potential for higher earnings and freedom from supervision. On the other hand, rates drove men to more strenuous effort, contributing to fatigue and injury."[46]

It's been speculated, and studies have suggested, that piece-rate wages may indeed compromise logger safety, as loggers are then inclined to work faster—and take greater risks—to cut wood more quickly so that at the end of the day they can be paid more. Piece-rate wages were also called "an invisible foreman" by one observer: "It is a vicious circle. The paper company pushes the contractor and the contractor drives his crew. The end result may be strained backs and a high number of injuries."[47]

Despite these assertions, empirical studies of the possible impacts of piece-rate wages on logging accidents suggest a lack of clarity on the matter. For example, based on the premise that piece-rate wages rewarded unsafe behavior by encouraging loggers to work more quickly and take additional risks, a Swedish study found that there were fewer and less severe logging accidents as that country transitioned from piece-rate to time-based wages in 1975. The author suggested that the "construction of the abolished piece-rate wages and the way of measuring individual productivity made the relation between actual work intensity and earnings obvious."[48] An earlier study from British Columbia, however, compared accident rates for salaried timber fallers with those working as "piecework-fallers." The author found that piece-rate wages were not associated with the rate of logging accidents. Perhaps not too surprising, a logger's age did explain accident rates among timber fallers, with younger, and presumably less experienced, loggers having the highest accident rates.[49]

The attorney in this case is likely trying to establish that the injured logger was paid by time to demonstrate that the logger was not compelled by his employer to rush his work or to take unnecessary risks because of the way he was paid.

DEFENSE ATTORNEY: Can you describe for me how your accident happened?

PLAINTIFF: That I don't know.

DEFENSE ATTORNEY: What's the last thing you remember prior to your accident?
PLAINTIFF: Last thing I remember I was on the ground.
DEFENSE ATTORNEY: Were you on the ground because something hit you?
PLAINTIFF: All I know I went to my knee and then to the ground.
DEFENSE ATTORNEY: What was the last thing you remember before your accident?
PLAINTIFF: I felled the tree and stepped back sideways.
DEFENSE ATTORNEY: So, the last thing you remember is stepping back.
PLAINTIFF: Getting away from the tree.

This chopper is apparently claiming that a tree struck him as he was initiating his retreat from a falling tree. It's unclear from the testimony whether the tree that struck him was one that he himself felled or another tree, perhaps a tree that was hung up in the tree he was felling and struck him from behind.

DEFENSE ATTORNEY: The last thing you remember is stepping back to get away from the tree.
PLAINTIFF: Yes.
DEFENSE ATTORNEY: And the tree was falling at that time?
PLAINTIFF: Yes.
DEFENSE ATTORNEY: After the accident, did you experience pain in any part of your body? Did any part of your body hurt?
PLAINTIFF: The bottom part of my body was hurting. Yes.
DEFENSE ATTORNEY: When you say you couldn't move, what parts of your body?
PLAINTIFF: Couldn't move.
DEFENSE ATTORNEY: What part couldn't you move?
PLAINTIFF: I couldn't move the bottom half of my body.
DEFENSE ATTORNEY: From your waist down or from your chest down?
PLAINTIFF: From my waist down.

The day-to-day physical rigors of woods work combined with the hazards of a sometimes harsh and unpredictable work environment combine to make logging, and especially timber falling, the most dangerous occupation on land in the United States. While post–World War II mechanization of the logging woods increased worker productivity, the increasingly competitive nature of the business continued to demand long hours with low pay and few employment benefits, if any. Worn down by the physical nature of the work, a hazardous forest environment, and the pressures to make it all pay, many loggers were done by 40. Pressures to produce and compete also eroded any sense of occupational community among the region's logging workforce. "As long as you have independent loggers, we won't stick together," asserted one northern forest logger.

Seemingly overnight, a new generation of postwar logging mechanization—chain saws, cable and grapple skidders, slashers, in-woods processors, forwarders, and feller-bunchers—emerged, replacing horses and river drives, crosscut saws, and axes. Logging's quirky jargon describing its basic work functions—chopping, choker-setting, and twitching and skidding—carried over to a new, more mechanized logging reality. Along the way, the hazards of the logging woods—hung trees, widowmakers, danger trees, and chainsaw kickback—evolved, too, with accidents occurring faster, more often, and at times more fatally.

In addition to pressures to produce more wood at less cost, methods of payment, such as piece-rate wages—"the invisible foreman"—often resulted in risk-taking behavior and poor decision making on the part of many loggers looking to increase their efficiency, ultimately contributing to an industry marked by a high rate of injury. A further complication was another type of invisible foreman, a management model that intentionally obscured the hierarchical relationship between the college-educated company forester and the so-called independent logging contractor, especially in Maine's cantankerous northern industrial forest.[50]

Part III

Unraveling

The people of rural Maine need good jobs with fringe benefits, and being employed as a logger by some of the forest products companies provided that. People who had enough time [as employees of a company] and lasted long enough were able to retire with some decency and some dignity, which you do not do working for a jobber. It just doesn't happen. But [providing employment benefits] costs you money. [Maintaining company logging crews] wasn't as cost effective as [hiring] a [logging] contractor.

—Michael Hillard, interview with Jim Pinkerton, June 15, 2006

Chapter Four
Cut Loose

How do we deal with the [independent logging] contractors? There is no limit to the variety of methods. Our two foresters supervise and cooperate with the contractors in the woods.

—Samuel G. Hall, manager for Draper Corporation, speaking at Session 3, "Problems of Contract Logging," at the Fifth Northeastern Logging Congress in North Conway, New Hampshire, in 1959

You say that these [logging] contractors are independent, and yet you tell them how you want the job done. How can they be independent then?

—Question posed by John Bork, assistant general logging superintendent for the Brown Company, Berlin, New Hampshire, and chair of Session 3, "Problems of Contract Logging," at the Fifth Northeastern Logging Congress in North Conway, New Hampshire, to Richard Waldron of Chadbourne Lumber Company, Bethel, Maine, in 1959

Logging labor and mechanization in the post–World War II northern forest was shifting, and a more distinct independent logging contractor profession began to emerge. Since that time there have been concerns expressed about the availability of men willing to engage in the hard work and low pay that characterized a logging sector becoming increasingly mechanized. According to one observer of woods work in the northern forest of eastern Canada, "Post-war tree harvesting was characterized by poorly paid, hard manual labor tethered to a seasonal workforce based primarily on semisubsistence agriculture. Given that the bulk of this labor force resorted to tree harvesting as a job of last resort during the agricultural 'downtime' of winter, foresters feared that this seasonal rural labor force would dissipate as agriculture became both more commercial and more mechanized, and as urban employment alternatives expanded in Eastern Canada." Describing the next phase of logging in the northern forest and the labor challenges it presented, he asserted: "Tree harvesting in this post-war era was completely dependent on horses to provide the motive power necessary to move logs and pulpwood from the stump to the roadside. The mechanization of agriculture meant a declining population of horses together with declining teamster skills as tractors became more commonplace . . . Because [mechanization] would convert tree harvesting from a seasonal, labor-intensive activity to a yearlong one with improved working conditions and wages, it was seen as essential in developing a response to the declining workforce."[1] During this period of transition from labor-intensive to mechanized logging, changes were coming to an industry that had to this point, in some parts of the region, especially in Maine's northern industrial forest, offered direct employment to much of its workforce. Toiling under "the eye of the stranger" would take on a new meaning, as a different logging woods authority—the industrial forester—tried to place some distance between himself and those who cut wood for his mill, while still holding onto a desire to manage a workforce of newly independent woods workers.[2] At the same time, loggers began to push back against the very independence to which many of them had aspired, initially oblivious to or unprepared for the responsibilities and costs associated with their newly acquired autonomy. Soon, both company foresters and independent logging contractors began to realize the challenges of trying to have it both ways.

BONDS AND JOBBERS

With erstwhile farmer-lumberjacks finding more lucrative livelihoods in the business of year-round commercial agriculture and World War II impacting the availability of logging labor, an observer of Maine's logging community noted that "the late 1930s and 1940s were years of crisis for the pulpwood industry, when woodland managers were forced to reevaluate the labor intensive, seasonal methods of pulpwood harvesting. The 1938 Fair Labor Standards Act and other New Deal legislation significantly raised labor costs by making employers responsible for the welfare of their employees."[3] One solution was increasing productivity through mechanization—essentially replacing manpower with machine power—or finding cheaper labor. Because of the labor shortage after the war, woodland managers for paper companies advocated for improved harvesting technology. Adapting to innovation in Maine's logging woods was slow, however, in part due to an increasing dependence on cheaper woods labor in the form of bonded Canadian woods workers, primarily from Québec. The result was a reliance on local "petty primary producers" and Canadian bonded labor in the state's more remote northern forest regions, with paper companies maintaining "strict control over the price, delivery, and quality of the wood coming into the mills"—a control that would be challenged by independent logging contractors some years later.[4] Indeed, a representative from Great Northern Paper Company in Maine's industrial forest declared in 1965 that the region's forest products businesses "have never experienced a serious shortage of woods labor," primarily because "Europeans and Canadians took care of our needs very well,"[5] referring to what the spokesperson described as "the bond program," partly implemented to provide a supplemental source of woods labor to the northern forest.

Writing in the same year, another Maine forest industry representative asserted that "the supply of wood in the northeast exceed(ed) demand" during the 1950s and early 1960s, suggesting that a lack of manpower was not a problem in the northern forest's logging woods of the mid-twentieth century. "People currently cutting have been and are held down as to quantity, shut off from producing, facing price cuts, etc. . . . The biggest mistake industry has made in the past is not allowing the good cutters and producers to develop maximum production." In addition, citizens of Maine's northern forest who were just entering the state's logging workforce had to "work alongside experienced, hard-working, high-earning, intelligent French Canadians," further discouraging the persistence of American loggers in Maine's logging woods.[6]

By World War II, "special programs for the importation of Canadian woods workers were first established . . . because there was a shortage of domestic workers to harvest pulpwood," according to one study.[7] The practice of allowing Canadian woodsmen to work in the United States continued after the war, and the Immigration and Nationality Act of 1952 allowed nonimmigrant "bonded" workers to be employed temporarily as woods workers. They were referred to as "bonded laborers," or simply "bonds," because logging employers were required to provide assurances that they would return to Canada at the end of their employment.[8] The 1986 Immigration Reform and Control Act separated the post–World

War II H-2 visa program into H-2A for agricultural workers and H-2B for nonagricultural workers. At the time, the H-2B program included woods workers, as trees and forest products were not considered perishable, agricultural commodities.

Not surprisingly, bonded woods workers "came primarily from depressed Québec farming parishes just across the border from northern Maine . . . They were indentured and permitted to work only for the employer who posted their bond. Operating from the old logging camps, they typically harvested five days a week and commuted home on weekends." Aside from Québec's persistent rural poverty, Canada's social system and proximity to Maine's more remote north woods have been posited as contributing to the influx of workers from Québec and their willingness, allegedly, to accept more marginal employment standards than their American counterparts. "Socialized medicine and socialized credit for logging equipment made it possible for Québec loggers to accept safety hazards and low-productivity logging chances that were not acceptable to local workers. By exporting its unemployment problem, the Québec government in effect hurt US loggers and subsidized the paper companies," claims that would echo for decades after.[9]

Bonded woods workers were fundamentally different from "visa woodsmen." According to a Great Northern Paper Company representative, "From 1940 on, [the bond program] worked very well," providing a stable supply of woods labor, until "around 1950 when somebody found out that it was more convenient to obtain an immigrant visa . . . Personally, I've not been very happy with the visa situation because it depleted our labor supply," the representative asserting that "only one out of every five visa holders continues to work in the woods. Since a visa gives our Canadian neighbors all the rights of a citizen except the vote, they've taken advantage of these rights to find year-round employment in other industries all over the New England states, and we certainly don't blame them."[10] In his 1960 synopsis of the program, another forest industry representative and proponent of the bonded labor program wrote that "bonded labor, unlike visa labor, has to be requisitioned by the employer. The employer must, under penalty of law, show preference for domestic labor whenever this type of labor is available. To requisition bonded labor, the employer must file a woods work order with the local Employment Commission."[11] According to some Maine loggers, arguments in favor of bonded labor didn't always hold up, becoming a basis for subsequent unrest.

Eventually, the logging community in Maine's more remote north woods comprised two populations—Maine citizens and cross-border Québec citizens—that were separated by nationality, culture, and language. A 1977 report, *Bonded Canadian Labor in New England's Logging Industry*, by the Massachusetts Agricultural Experiment Station found that only three logging businesses employed Canadian labor outside of the Maine's border counties with Québec.[12] At the same time, Maine loggers began to transition from being employees of the large paper companies to independent logging contractors, or "jobbers." Many employees jumped at the opportunity to become independent, to be their own bosses, thinking that their newfound sense of frontier individualism and self-determination would translate into greater opportunity and less reliance on logging camp life. However, many newly independent loggers were unaware of what lay ahead and the implications of no longer being employees of the mills. According to one source, "thousands of loggers appear to have been happy to exchange the life of wage earners in isolated logging camps for that of self-employed commuters" but were unaware of the challenges associated with running a business, of being their own boss, in an environment where the mills and their company foresters often continued to call the shots in the state's logging woods.[13] About the transition from logging employee to independent logger, a Maine legislator, who was a log hauler from Aroostook County, asserted in 2019: "One day he's an employee, and the next day he was a contractor. I mean, it just happened like that . . . Most people, myself included, weren't really sophisticated enough to know what the downfall of that would be."[14] The legislator expressed a sense of victimization that didn't always play well among other loggers in the northern forest.

Whether these northern Maine loggers were truly victims or perpetrators of their own demise as independent logging contractors has been rarely addressed, the potential political ramifications likely discouraging an open discourse on the matter. In any case, not all northern forest loggers have been sympathetic to the plight of independent logging contractors in Maine's north woods who worked as service providers to the large paper companies in the state. During a group meeting of Vermont loggers in 2000, one participant stated emphatically and to general approbation from other loggers in the room: "I have no sympathy for those loggers in northern Maine who went to bed with the SFI companies and have invested in too much logging production capacity based on promises from the large landowners," a reference to the Sustainable Forestry Initiative, an attempt by the forestry sector to establish its own industry-approved and monitored guidelines for forest management and harvesting that have implications for forest stewardship, best management practices, and the chain of custody of forest products.[15] If nothing else, the Vermont logger's comment emphasizes the complexity and diversity of the logging community of the northern forest and cautions against broad assertions that attempt to characterize the logging community of Maine or the region. It also suggests that aspects of the logging community in northern Maine may be considered outliers in the logging woods of the broader northern forest region. Indeed, the logging woods of Maine's industrial forest, it may be argued, represent a place of antagonistic mutualism, a complex ecosystem of interdependent, but often contentious, relationships between increasingly mechanized and capitalized logging businesses owned by nominally independent logging contractors and large, regionally dominant wood-consuming mills and industrial forest owners.

As the notion of an independent logging contractor in the northern forest evolved, so did an organizational hierarchy of woods work and jargon. According to one logging historian, "When lumberjacks, previously working for day-wages, shifted to 'piece' work—i.e., payment by the cord for cutting four-foot pulpwood—they earned about $1.25 per day . . . For years the companies established the camps and hired and paid the men. Nowadays they 'job' out the season's cut: they hire various men in whom they have confidence to cut and 'land' the number of cords they figure they will need . . . they prefer the more grandiose title of 'contractor' to that of 'jobber' . . . These contractors in turn let out parts of their contract to lesser jobbers. It is a curious fact that a jobber can cut and land wood, and drive it, too, more cheaply than a company can. The company of course helps the contractors and jobbers with equipment or even loans."[16] While company logging crews were something of an insurance policy for mills that guaranteed a certain supply of raw material, these crews were salaried and had company benefits like other company employees. Ultimately, the motivation for mills to divest themselves of their company logging crews was to reduce costs that would normally accrue to employees, including those associated with worker safety and health care, as well as the fixed and variable costs associated with owning and operating logging equipment. That some companies would continue to supervise newly independent logging contractors and help them with equipment loans would soon become a source of tension around what it meant to be an independent woods worker and logging business owner.

NOMINAL INDEPENDENCE

A post–World War II shift from company logging crews to independent logging businesses was on. "Company crews were gradually laid off and loggers required to become independent contractors . . . supplying their own equipment and working for piece-rate pay," according to one writer. "In 1945, over 70 percent of timber was still harvested by company crews; in 1956, it was less than 50 percent and by 1968, less than 20 percent."[17] In today's northern forest there are very few loggers who work directly as employees of forest products companies, the transition from company crews to independent logging contractors having been mostly completed by the 1980s, although one Maine company, Scott Paper, held onto some of its company logger crews through the 1990s.

However, the shift to independence was more challenging than many loggers had anticipated, and some in the region continued to feel the impacts. Almost overnight, logging employees—knowledgeable and experienced in felling and skidding—became logging entrepreneurs, now required not only to cut trees and move wood but also to sustain businesses, including negotiating contracts with wood-consuming mills. The widely divergent skill sets associated with being both a logger and a businessman were not necessarily well developed in all former employees of the mills. One observer of the history of logging in New York's Adirondacks noted that "the mechanical revolution in logging during and after WW II produced many economic and social changes. The old-time lumberjack and his lumber camp home had largely disappeared . . . [new equipment] required an investment of thousands of dollars; operating costs were high; markets were sometimes uncertain; state laws were often unfavorable; and other difficulties frequently arose."[18] Among the "other difficulties" facing new independents were competition from

Canadian bonded and visa woods workers, the procurement of stumpage and favorable logging chances, and managing the nuances of contract logging, including their own logging employees.

On the subject of the high cost of the industry's evolving post–World War II mechanization—from skidding with horses to skidding with tractors—one northern forest procurement manager offered observations in 1959 that would certainly ring true for the more mechanized logging businesses of the early twenty-first century: "Skidding small logs from thinned stands in our mountains awaits a method better than horses. Tractor costs have risen rapidly, yet our textile directed markets do not allow increased log costs. An obvious mistake which can be made by company logging lies in overcapitalizing in expensive machines to mechanize operations. Our wood requirements are relatively small, two million of maple at Beebe River [New Hampshire] annually, plus an equal volume of softwood sawlogs. There is no room for heavy burden in depreciation."[19] Indeed, equipment depreciation and replacement costs were likely rarely accounted for by the newly independent business owners, their primary focus being on managing their day-to-day operating costs—fuel, lubrication, replacement of tires and tracks, and equipment repairs and maintenance—that were viewed as more immediate and tangible. Although woods production was increasing with logging mechanization, so were costs: and the race between the two was on. In an attempt to come out ahead, many independent loggers worked longer hours, which took a toll on both man and machine. Some chose to avoid the costs of having employees, of training and initially losing money on logging greenhorns, only to have them move on to better-paying, more stable year-round work in another field. Thus they decided to go it alone as owner-operators, increasing their isolation, compromising their safety, decreasing logging efficiency (the skidder would sit while the owner-operator would chop a twitch—a payload of cut and delimbed trees—to drag to the landing), and contributing to a further breakdown in any sense of a logging "community."

From the perspective of the forest products industry, the motivation for the independent logging contractor arrangement was clear. Logging mechanization and contract logging arrived in the region's woods at about the same time, and the transition was often painful. But not for consuming mills, which now bore few, if any, direct costs associated with felling trees and moving logs. A key outcome of this transition from company crews to independent logging businesses was that, as independent contractors, "there is no Occupational Health and Safety Administration Protections, no workers' compensation requirement, no health plan, and no withholding for FICA or Medicare," according to a 1999 Department of Labor study cited by the author of a *Maine Times* article titled "Hard Times in Irving's Woods."[20] A study published a few years later confirmed that few northern forest logging employees were covered by logging company–sponsored health insurance.

Addressing the challenges associated with the then still relatively new independent logging contractor phenomenon, a series of three articles from the Fifth Northeastern Logging Congress held in North Conway, New Hampshire, in 1959 was published in the *Northeastern Logger* trade magazine under the general heading "Problems with Contract Logging." Working with independent logging business owners was discussed from the perspectives of northern forest mill and procurement managers. The articles were published during a period of abrupt transitions: by the end of World War II, according to one estimate, so-called company crews harvested more than 70 percent of the pulpwood, and by the mid-1950s that proportion was less than 50 percent and less than 20 percent by 1968.[21] The era of the independent logging contractor in the northern forest had taken hold, but not without controversy around what it meant to be truly "independent."

One presenter at the congress, Samuel Hall, a manager for Draper Corporation's Beebe River, New Hampshire, bobbin manufacturing facility, stated that "logging is a problem in cost control, and control in the woods is difficult. Contract logging solves the control problem by dividing up the risk, by concentrating the effort and by getting units delivered at established prices."[22] The risk that was once the sole responsibility of the paper companies was now being shared with independent logging contractors. Indeed, the risk associated with the costs of logging equipment ownership and operation, coupled with the unpredictable nature of the logging woods environment, including topography, weather, and ground conditions, now became the sole possession of the logging contractor. Describing wood procurement for Draper,[23] an operation that manufactured maple loom bobbins, for which raw material was estimated to be about a third of the total product cost, Hall pointed out that the company's "current log supply coming from 80 suppliers last season consisted of 75 percent purchase logs, 9 percent company operated logs, and 16 percent contract logging logs," accentuating the post–World War II shift from company logging crews to independent logging contractors and petty producers.[24]

Clarifying the mill's relationship with logging contractors, Hall continued: "The least complicated and one of the most satisfactory arrangements is to sell the stumpage to the operator. He then delivers to our mills the species we want, marketing the remaining grades and species to his satisfaction . . . One of the reasons why we like to sell stumpage is that it divides the liability." As to what type of contractual arrangement Draper had with independent logging contractors, Hall stated: "I think that it is a peculiarity of our region to be able to do a considerable volume of business without any written contracts whatsoever." Selling stumpage also helped distance the stumpage seller from risks associated with logging, including accidents that inevitably occurred in the logging woods and the legal liability associated with logging-related injuries or poor workmanship and carelessness that could result in impacts to the forest environment, such as residual stand damage and impacts to water quality. Elaborating on the advantages of selling stumpage to logging contractors, only to buy it back in log form, Hall wrote: "I think it further establishes him as a completely independent businessman, insofar as wage and hourly regulations are concerned. And it's a convenient way of doing business," ambiguities around Draper's relationship with logging contractors—legal and otherwise—notwithstanding. As to how independent logging contractors are "dealt with," Hall explained that Beebe River's foresters "supervise and cooperate with the contractors in the woods," a relationship that would cloud issues related to both a logging contractor's independence and liability for logging accidents, as we'll see later.[25]

During the same 1959 discussion among mill representatives, Robert Moodie, a woods operations manager for Scott Paper in Winslow, Maine, weighed in on the role of contract logging in the pulpwood industry, observing that "'contract logging' in the northeast pulpwood business means a method of logging where an independent individual or company contracts to produce or deliver to the wood user a certain quantity of wood for a specified price." Elaborating further, he explained that "in the broadest interpretation 'contract logging' is the method where the whole logging job, from stump to mill, is undertaken by a responsible individual, who does this with his own resources and under his own direction. As opposed to this, we have the company logging method where the wood-using company manages and performs the logging job itself."[26]

Not long after Moodie made these assertions, company crews were nearly nonexistent in the northern forest, loggers having been cut loose as employees of wood-consuming mills to become independent logging contractors. The word *independent*, however, was open to broad interpretation. Truly independent loggers were stumpage buyers, either paying up front for all trees to be harvested during a timber sale ("lump sum") or paying the landowner for timber that is harvested as it is cut and delivered ("unit price" or "pay as cut"). Often, however, in the industrial forest of northern Maine independent contractors were essentially service providers, working for industrial forestland owners and mills on service contracts, sometimes referred to as CLS agreements—contracts for logging services. Loggers who were contracted as service providers did not generally purchase timber on the stump—"stumpage"—but rather provided a service to the owner of the stumpage and were essentially paid piece-rate wages for their work.

But piece-rate payment systems in logging have been criticized beyond their implications for logger safety, as discussed earlier. Suggesting a lack of flexibility in the ways that piece-rate wages were set, in 1961, as the creation of a logging workforce increasingly dominated by independents was picking up steam, one writer asserted that while rates were often simplistically established for "average" logging conditions, they didn't adequately account for the variability in logging chances, such as topography, amount and size of trees to harvest, skidding and trucking distances, and the condition of forest roads and trails, which could have a significant impact on logging costs.[27] Years later, a former Maine logger offered another criticism of piece-rate wage systems, one that was tied more closely to the feeling among some loggers that they were manipulated by the mills: "If you work faster, they [the mills] cut the rate . . . [and] you couldn't push back. All you could do was work as hard as you could. And so the guys in the woods were squeezed."[28] Others have criticized the lack of a fair scale for delivered wood, which has often led to acrimony and distrust between loggers and wood-consuming mills.[29]

Moodie went on to outline advantages of contract logging to wood-consuming mills, including:

> "The true contractor on a firm price will absorb the losses for adverse weather, labor shortages, unanticipated difficulties, etc., in the same way that he takes advantage of favorable factors";
>
> "The contractor relieves the company of large capital investment for logging equipment. He also takes the losses of depreciation and possible obsolescence of machinery resulting from idle periods"; and
>
> "Competition for the lowest possible price can be obtained by bidding between two or more contractors," with the wood-consuming mill gaining "in bargaining strength by the fact that there is more than one applicant for the job."[30]

Disadvantages to contract logging included a loss of flexibility "in adapting operated volumes to suit changing mill requirements" and the loss of some efficiencies "such as large-scale purchasing of supplies." Despite these drawbacks, Moodie concluded that "contract logging is necessary in many ways to keep our wood costs competitive," those costs having been transferred to a developing workforce of nominally independent logging contractors and petty producers.[31] However, the delineation of work functions between wood procurement foresters and loggers had not been fully negotiated, often resulting in a lack of clarity around the operational meanings of terms such as *independent logger* and *logging employee* that have persisted in today's logging woods.

When asked about the distinctions between loggers as independent contractors and employees of a wood consuming mill, the conversation among the mill representatives at the Northeastern Logging Congress went underground. Referencing a case in Oregon, one participant stated that even though an independent logging contractor may have compensation insurance, for example, "unless he had complete control over the operation he was considered to be an employee, in the eyes of the court." Apparently becoming uncomfortable with the flow of the conversation and its implications for wood-consuming mills around the assertion of "independence" among logging businesses, the chairman of the group discussion weighed in: "I think it has been rather obvious by the few questions that have been asked, that everyone would like to ask more questions about this particular subject, but that everyone is afraid to ask such questions. This subject of dependence or independence, so far as the contractors are concerned, is a very moot point. Everyone seems to be saying 'I'm not going to stick my neck out at a gathering of this nature.' I agree that there needs to be a clarification, and yet many people are afraid of clarification." To which another participant declared, "The remainder of the discussion at this session will be off the record," clearly signaling discomfort with further discussion about the evolving relationship between independent logging contractors and wood-consuming mills that the industry had brought upon itself.[32]

Dismissing the relevance of dependence versus independence in the logging woods as being "very moot" underestimated the challenges ahead. Indeed, one cocksure congress participant, Richard Waldron from Chadbourne Lumber Company of Bethel, Maine, weighed in, emphatically asserting in capital letters: "YOU MUST SUPERVISE ANY AND ALL JOBS," referring to the responsibility of wood-consuming mills in their relationship with independent logging contractors. "It is only natural for any contractor to want to grow and enlarge his crew and equipment, but in New England the small, efficient contractor is what we want . . . 'Keep him small and on the ball,'" he declared, suggesting a form of control over independent logging contractors that constrained their ability to grow. Spreading a mill's wood procurement requirements over many logging contractors and petty producers, forcing them to stay small and compete against one another for contracts, accomplished this goal. When Waldron was asked how he knew his contractors' operating costs so that he could make more informed decisions on contracts, he replied, "I have access to any contractor's books." This was followed by a key question from one skeptical participant: "You say that these contractors are independent, and yet you tell them how you want the job done. How can they be independent then?" Waldron replied that he does not tell the logging contractor how to log but rather "what you want done, and pretty much how you want it done," citing preserving forest regeneration during the logging operation as an example. The questioner pressed him further:

QUESTIONER: "I don't see how you can do all this, and still claim that your contractors are independent."

WALDRON: "When he finishes a job I have no hold on him, until he accepts another job. He is free to go to work for someone else. These contracts are all verbal, except that I usually make a few notes, so I don't forget what we have decided."[33]

The implication was that it was acceptable for a forester or other agent of a forest or stumpage owner to direct an independent logging contractor as if he were an employee while he was working but not to give direction about with whom he contracted. However, especially in some remote regions of northern Maine, wood-consuming mills represent a monopsony, where a single wood purchaser substantially controls the market, wielding significant power over wood producers. Under such market realities, there was little, if any, hope for a logging contractor to find woods work or a realistic market for his wood beyond the one company that dominated the region. That a manager for Chadbourne Lumber Company had access to independent logging contractors' books indicated a level of oversight and knowledge that would transcend that of a contractor-contractee relationship, at least as it was suggested by his questioner. The new "nominally independent" logging contractor, disadvantaged by a power imbalance with the mills with whom he contracted, had emerged in the northern forest, defining generations of acrimony to come in Maine.

Decades later, in a 2000 article titled "Hard Times in Irving's Woods," one northern Maine logger, complaining about the treatment of loggers by industrial forestland owner and forest products manufacturer J. D. Irving Company, asserted that "you're a contractor now if you do just one part of the work. They tell you what to do at every turn. I don't feel independent. It's a joke . . . We have the independence to only say yes or no to a contract. We're controlled. We should be employees as long as they have this kind of control." For its part, an Irving representative explained that "paper and lumber markets are more competitive than ever, and all the players are producing high quality and lower prices. You have to meet them or go by the boards," which indicated that neither side was inclined toward a more cooperative, mutualistic relationship.[34]

One manager of a forest products company in nearby New Hampshire who participated in the 1959 Northeastern Logging Congress chronicled by the *Northeastern Logger* claimed that "dealing with all log suppliers has its transient difficulties, its chuckles and satisfactions, however, the loggers are our best friends and we intend to keep it that way."[35] Not all loggers would likely have agreed, whatever chuckles may have existed giving way to frustration with their new status as "independents" in an increasingly global forest products market.

Indeed, adding some depth to the matter of logger independence, the often-adversarial relationship between independent loggers and foresters emerged during a 2005 group discussion of eleven Québécois loggers who worked in Maine, held in St-George-de-Beauce, Québec. Some group participants asserted that a logging capacity crisis was created by Maine's procurement foresters seeking to drive down prices for raw material by encouraging more suppliers to enter the wood market, keeping them "small and on the ball," to use the language of the Chadbourne Lumber Company's manager almost a half-century earlier. Others criticized foresters for their lack of appreciation for the knowledge and skills required of professional loggers. "Loggers have underutilized practical knowledge. We're not taken seriously by foresters." Despite a sense of some improvement, "still there isn't much trust in loggers." Echoing a common refrain on both sides of the border, one logger added, "The forester who paints the trees doesn't have an idea about how you will be driving through the woods. It doesn't make any sense to pay somebody to do that. We could do the planning ourselves." Indeed, the tension between some foresters and loggers described by these Québécois woods workers represents an undercurrent of distrust and disdain not uncommon in the region's logging woods. One writer described the tension between the two professions as a "clash of occupational cultures"—the forester is trained to "follow and enforce the rules," whereas the logger is "socialized to disdain them."[36] But it is probably more complicated than so described. Most foresters, trained in the science and practice of their profession, have never been loggers and yet are often charged with "supervising" them during harvest operations. Despite this disconnect, loggers contended that some foresters assumed an authoritative, paternalistic approach toward loggers, who quite reasonably viewed themselves as independent and more knowledgeable and skilled in their own profession than foresters.[37]

Not everyone agreed with the industrial forest landowners and wood-consuming mills about the region's newly independent logging contractors. Referring to what he called the "corporate domination of the forest economy," one observer viewed things differently some forty years after the 1959 meeting among northern forest mill representatives that was documented in the *Northeastern Logger*: "Paper companies have depressed stumpage prices paid to non-industrial forest owners through an evolving mix of monopoly power, collusion, and tactical harvesting on their own lands. They have countered upward wage pressure at various times by importing Canadian labor, converting from fixed wages to piece-rate, forcing hired crew to become contractors, and promoting labor-saving techniques." Describing Maine's forest history as "a class-based plot," he asserted further that loggers in northern Maine "have been victimized by chauvinistic hostility between Yankee and Franco-American interests and by anachronistic aspirations of self-reliance and frontier individualism."[38]

But his assertion that "from 1945 to the mid-1980s, most timber was cut by small, nominally self-employed crews that worked logging 'chances' with little supervision" cannot be applied universally to loggers in Maine or the northern forest region during that period. Although many logging contractors may well have been "nominally self-employed," it was not uncommon for even these "independents" to receive direction from foresters representing mills or landowners. Referring to "the fictions of independence" among Maine's loggers of the period, he later observed that "a species of frontier individualism, manifested in the desire for (nominal) self-employment, appears to characterize many loggers . . . [T]here is compelling evidence to support the conclusion that the system has also entailed exploitation of loggers, degradation of forest ecosystems and erosion of traditional rural ways of life."[39] Soon, descriptions of logging as a traditional rural way of life would not necessarily resonate well in a state struggling to achieve greater economic relevance by investing in technology, innovation, and an urban economy, as we'll see later.

A COMPLICATED INDEPENDENCE

In his 1974 book *The Paper Plantation*, William Osborn, referring to logging as "one of the lowliest and least desirable occupations in Maine," offered the following generalization: "The pulpwood procurement system has driven the logger into debt and deprived him of even the most rudimentary of social benefits."[40] Discussing contract logging in Maine, Bowdoin College economist David Vail described logging in terms of "'human obsolescence,' manifested in poverty earnings, high injury rates, middle-age 'burn-out' and rising unemployment . . . [A] basic historical fact . . . is the weakness of ecological consciousness (greenness) and class-based organisations [*sic*] (redness) in the Maine woods,"[41] the latter an interesting, but rather sweeping and unsubstantiated, social commentary about a much more complex place than a single "Maine woods." About a decade later, researchers at the University of Maine found that Maine loggers chose their profession, rather than being in any way forced or coerced into it. According to a study of hundreds of the state's logging contractors and logging employees, individuals were so employed because of a number of job attributes, especially a sense of independence, working outdoors, and the belief that logging was healthy work, the "human obsolescence" asserted by Osborn and others notwithstanding.[42]

Despite studies that have repeatedly shown that a vast majority of loggers work in the northern forest because of the sense of independence associated with woods work, issues related to the industry's transition away from company logging crews presaged tensions between newly independent contract loggers and the mills. A fundamental problem with the new independent logging contractor paradigm was that "the small independent (woods) worker does not have the capital sufficient to finance his business," according to one Maine forestry sector representative interviewed for the 1980 documentary *Cut and Run*. During that period, *Cut and Run* estimated that of 7,000 loggers working in the Maine woods, 1,700 were employed directly by paper companies; the rest were independent woodsmen. Agreeing with participants in the group discussion of wood procurement foresters some twenty years earlier, "by and large it is much cheaper [for the companies] to receive wood from independent woodsmen than to cut it themselves," asserted the narrator.[43] This didn't always work out well for independent logging contractors, who, over time and as logging mechanization took hold in the region, complained about poor logging chances, inadequate delivered log prices, and wood quotas that were too low and too unpredictable. A Vermont logger who participated in a group meeting in 2000 about the state of the logging profession offered this comment: "When you have that much invested in your business and you can get shut off [by the mill] with a message on your answering machine . . . ," followed by another who asserted that "it's one thing to be a logger; it's another thing to make a profit," highlighting the stark differences between being a competent logger and being a successful businessman.

In *The Paper Plantation*, Osborn described independent logging contractors this way: "Since they are not the employees of the mills they supply, the independents purchase and own the machinery—the chain saws, skidders, bulldozers, and trucks necessary to a modern logging operation . . . Although he is generally referred to as an "independent" contractor, the small pulpwood operator enjoys little freedom or flexibility in his business and is in fact completely dependent on the large paper companies. His activity is tightly controlled by a series of one-sided contracts. Before an operator can go into business he must obtain both a stumpage permit from a landowner or owner of timber rights and a contract to deliver wood to the pulp mill."[44] Adding to the dilemma, newly independent loggers interested in purchasing stumpage, anticipating the prospect of making some profit from doing so while being able to truly behave as independent business owners, are often challenged by competition in bidding for stumpage from well-funded mill procurement organizations that are capable of driving up the cost of stumpage, thereby eliminating, or at least severely challenging, competition from logging business owners. In addition, maintaining a large pool of raw material suppliers allows some mills and procurement foresters to either play one logger against another when negotiating prices or completely eliminate some from the conversation, confident that other suppliers will take their place.

Almost a decade after the 1959 Northeastern Logging Congress session that addressed the burgeoning independent logging contractor paradigm in the region, one observer of Maine's logging woods suggested that "there is probably little question that labor problems will increase in the years to come."[45] He would not have to wait long.

As with any business that provides professional services, independent logging contractors competed against one another for wood contracts and during stumpage sales. The difference was the unpredictable nature of logging, work that took place in an environment of great uncertainty related to weather, challenging terrain, shifting markets, and unreliable woods labor. Some northern forest mill managers admitted to exerting supervisory control over nominally independent logging businesses while encouraging an overbuilt workforce of petty producers so as to drive down their raw material costs—keeping loggers "small and on the ball."

That a single wood-consuming mill might also be "the only show in town"—a monopsony—as was the case in some remote regions of Maine's northern forest, shifted negotiating power to the mills, which could play one logging contractor against another, as described by Scott Paper's woods operations manager during the Northeastern Logging Congress in 1959, a conversation that eventually went underground because of its provocative take on the relationship between mills and newly independent logging contractors. Any notion of a regional community of logging professionals was being compromised by a power imbalance with wood-consuming mills and an increasing sense of resentment toward bonded woods labor. All players in the system were in their own ways complicit: loggers naively embracing a newfound independence without considering the ramifications, then claiming to be victims; mills prioritizing short-term profit margins over long-term wood supply and the health of the rural workers, families, and communities on which they depended.

Despite a real sense of urgency in the northern forest, the two sides chose to engage in distracting hyperbole, with little effort by loggers and wood-consuming mills to abandon their adversarial relationship and instead cooperatively address the emerging challenges faced by the state's forest products industry: woods labor controversies, the sustainability of logging businesses, questionable procurement practices, and global competition. The contentious relationship between nominally independent logging contractors and the mills was becoming more clearly defined, with cooperative solutions never fully negotiated. Victimization and greed prevailed over collaboration and community. Maine's most important industry had begun to consume itself.

Chapter Five
Pushing Back

What's happening here is that the Canadians are much less of a hassle to have as employees because they demand no benefits, and I think they have been known at times to work for less than the scheduled rate, although I can't say for sure whether they're doing that now.

—An unemployed Maine logger,
interviewed in Richard Searls's *Cut and Run*, 1980

I wouldn't want an American to say that a Canadian comes over here to take my job. I am ready to work any time with an American, and I'm sure I can work with him and make a good team, because an American is capable of working hard.

—A Canadian logger working in Maine,
interviewed in Richard Searls's *Cut and Run*, 1980

Maine's acrimonious logging woods history is complicated, with roots extending back to the region's seasonal farmer-lumbermen, poor employment and education opportunities in much of rural southern Québec, a lingering sense of nativism toward Maine's Québécois citizens and guest workers, a newly independent Maine logging workforce, and the challenges of maintaining a globally competitive forest industry in the state's remote northern forest. A constant source of tension between independent loggers and mills in northern Maine has been the prices set by mills for delivered wood that were, according to some observers, "kept low by the paper companies that control the pulpwood market by cutting their own lands with bonded French-Canadian labor."[1] "The economics kept getting worse and the prices didn't go up very much . . . We were trying to get a little more money so we could afford to operate," according to one logger. At the same time, the H-2 program "guaranteed the job for six months when they [Canadians] come over here; we weren't guaranteed for one day." Regarding wood contracts offered by the paper companies, he asserted, "You take it or you leave it."[2] There were also complaints from Maine loggers that they weren't being assigned productive logging opportunities and that the best chances were going to Canadians, who were hired in ways that walked around existing labor laws. "We had so many examples of Americans who were put in very poor chances and weren't earning much money," according to one source.[3]

Although little or no systematically gathered evidence supports the assertion that Québécois loggers received preferential treatment in the assignment of logging opportunities, additional anecdotal evidence showing a bias in favor of Canadian logging crews includes a comment by a graduate of a logger training program in Maine during the 1960s, who claimed that "there was no future [in woods work] because the Canadians get the best chances." However, perhaps more objective is the open-ended perspective of the Maine Department of Conservation's report titled *Maine Woods Labor Study, 1984–85*, which stated that "one important MWA contention that has not been addressed, and upon which a variety of opinions are held, is whether or not Canadian laborers continue to get better 'cutting chances' than Maine loggers in the northern part of the state," referring to the Maine Woodsmen's Association, a labor organization formed in 1975 to push back against the paper companies. "Company officials assert that this practice has ended," the authors wrote, "while some independent loggers suggest that it continues," implying that some favoritism toward Canadian woodsmen did exist, at least in the opinion of some unnamed "company officials."[4]

A PRELUDE

Wearisome argumentation about whether Canadian loggers outworked Mainers aside, why would the paper companies want to hire Canadians, given the friction it caused with local Mainers and the contention that it placed companies on the wrong side of federal labor laws? "Historically and factually, foreign workers were less likely to complain to the authorities, you know, about violations of the law or poor or dangerous conditions," according to one source. Canadians were considered a "more docile workforce," at least in the eyes of the paper companies, he maintained. There were also tax benefits to hiring Canadians. "[Canadians] earned as much as the Americans, but they took home more,"[5] in part because of favorable exchange rates at the time.

Perspectives on the matter of Canadians working in Maine's logging woods varied widely, however. In its assessment of the bonded labor program, one report concluded that there was not an adequate domestic workforce to satisfy demand for loggers due to a number of attributes generally associated with woods work, including seasonality of the work; an "unsubstantiated" sense that wages in the woods were low, allowing that "higher wages would attract a greater number of domestic workers into timber harvesting;"[6] and the isolated nature of woods work and logging camp life. On the latter point, in 1974 Osborn asserted that "this remoteness, it is said, makes it impossible to get Maine men to work in the woods because they are less willing than Canadians to leave their families and homes overnight while they stay at company camps. The isolation of the cutting operations is overplayed by the industry, which would rather blame geographical distance than working conditions for the shortage of American cutters in the woods." According to Osborn, "It is the low prevailing wage, which many Canadian woodsmen will accept, but Americans will not, that is chiefly responsible for noncompliance with the condition that the bonded program not adversely affect domestic workers. The use of Canadian labor to harvest their lands, at wages and under conditions domestics will not accept, enables the companies to create a surplus of pulpwood for the mills, and to keep the price of purchased pulpwood down."[7]

The isolation of logging camp life wasn't the only impediment for Mainers. Language was also a hurdle, but one that may have more often challenged Maine's English-speaking loggers. According to the 1977 *Report on Bonded Canadian Labor in New England's Logging Industry*, "in logging camps dominated by French Canadians, the non-French-speaking US worker is likely to lead a lonely existence . . . [T]here is little question that it is a factor in deterring English-speaking Americans from seeking work where French is the primary language."[8] Based primarily on anecdotal evidence, the report cited several additional, now familiar factors that discouraged people from woods work: the physical rigor and danger of logging; the lack of fringe benefits, job security, and opportunity for advancement; and the low occupational prestige associated with logging—factors that were strongly echoed decades later in a 2009 University of Maine study of the state's logging profession.[9]

In addition, there was speculation that paper companies did not want local Maine loggers to work their woodland and, over time, develop settlements nearby that would evolve into communities that companies would have to support. “We reasoned that the Canadians were there to cut wood and go home. They weren’t going to encroach and move closer and closer to the wilderness, form towns, open schools, have police departments, tax the paper companies We thought that the desire to have Canadians was motivated by their not wanting to have settlement close to their produce. But that’s not something that we were ever able to demonstrate.”[10] In short, the contention goes, the paper companies had no interest in the corporate responsibilities associated with community building and the formation and stewardship of company towns, an attitude that seemed to be confirmed in the ensuing decades as paper companies abandoned many of Maine’s mill towns.

Although there had been scattered skirmishes around these issues in Maine’s north woods in the past, these efforts were short lived and lacked cohesion. However, matters came to a head in the 1970s, as some of Maine’s independent loggers began to push back against the paper companies. Given their sense of frontier independence and individualism, isolated workplaces, and the necessity of working many long and exhausting hours to make a living, loggers attempted something they weren’t particularly adept at. They organized.

Maine’s loggers began to agitate more emphatically against the paper companies around what they considered to be unfair and unlawful practices. “In the spring of 1970 a small group of independent operators decided to challenge the paper industry’s iron-fisted control of the pulpwood markets,” forming the Maine Pulpwood Producers Association (MPPA) “to boost the price of pulpwood” by five dollars per cord, threatening the supply of wood to the mills by “staging an event to dramatize their plight.”[11] The MPPA initiated its action by setting fire to a couple of cords of wood near International Paper Company’s mill in Jay, hoping to get the attention of the pulpwood industry and raise public awareness of the loggers’ plight.

Having failed to bring any changes to the way they were treated, many of the loggers decided to boycott two Maine pulp mills—International Paper and Oxford Paper—withholding pulp logs and encouraging other loggers to do the same. However, although the holdout curtailed delivery of an estimated two-thirds of the pulpwood normally delivered by independent logging contractors, the rather weak collective action was largely ignored by the industry, and the pulpwood burn appeared to backfire, serving only to anger a more environmentally aware Maine citizenry concerned about the impacts of a public wood burning on local air quality. While acknowledging that the loggers participating in the strike suffered "a disparity of bargaining position," the Franklin County Superior Court judge found the loggers' action in violation of state law against price fixing and a violation of the 1890 Sherman Antitrust Act, effectively bringing an end to the matter and taking the wind out of the sails of the MPPA.

Maine's troubles weren't finished, however, and the state's logging woods became a dangerous place to be if you were a Canadian logger, as more militant U.S. loggers switched their attention from the paper companies to woods workers from north of the border. The *New York Times* wrote this about the continuing tensions: "There is trouble in the Maine woods. Festering for years, it pits the poor, traditionally independent American woodsmen against the giant pulp and paper companies that dominate this state's land, economy and politics. The immediate issue in this remote area is the use of bonded Canadian woodcutters—which the Americans charge keeps wages down."[12] Despite its failed attempt to increase pulpwood rates in the short term, there was longer-term value in the uprising by the MPPA. According to a 1986 University of Maine master's thesis: "The Maine Pulpwood Producers boycott fit into the pattern of labor agitation in the woods; it was localized, fairly spontaneous, and short lived. However, the 1970 strike marked the first time that independent contractors had worked together in an organized manner. Although they were defeated, the small group of jobbers raised new expectations. The years immediately following the MPPA boycott were marked by isolated but persistent labor militancy. In the Allagash region between 1972 and 1974, protests over the use of Canadian cutters and contractors resulted in seven woods fires and an estimated $150,000 to $175,000 in damage to camps and equipment." He concluded that as tensions spread in the state's logging woods, "it would not be long before woodsmen banded together in numbers hitherto unseen in Maine."[13]

The paper companies, too, sensed trouble ahead in the state's logging woods. On June 20, 1975, an entry in the newsletter of the Great Northern Paper Company (GNP) referred to a "growing militancy among cutters." Maine loggers "want something done about the fact that bonded Canadian laborers are cutting wood in Maine. They want a change in the practice of mills cutting off purchases without notice," apparently acknowledging a practice that was a source of some discontent among the state's woodsmen. The following week, the GNP newsletter stated that "there is no doubt the cutters are going to fight to get bonded Canadian laborers out of the Maine woods."[14] Indeed, a couple of months later, "American woodsmen seized Camp 106 and sent the rival Canadian workers packing . . . The disgruntled [American] woodsmen have sometimes taken their fight into the forest by wrecking equipment or setting fire to logging camps or patches of woods," according to the *New York Times*.[15] More conflict was on the way and the state's industrial north woods would never be the same.

UNRAVELING

Opening with the crackling of Robert Charlebois and Louise Forestier from the radio of a 1970s GMC pickup truck carrying Québécois loggers crossing the border from Saint-Théophile to Jackman, Maine, at 5:00 a.m. on a cold Monday morning, the documentary *Cut and Run* highlights mid-1970s tensions between American and Québécois woods workers and the companies that hired them. The film was controversial. "In '80, when 'Cut and Run' was released, they [Maine's paper industry and Forest Products Council] sued not to get it shown," according to an MWA organizer interviewed in 2004. "So there was [*sic*] a lot of enemies of 'Cut and Run,' but we showed it hundreds of times."[16] Another MWA organizer claimed that not even Maine Public Television would play the film when it was released. "They were pretty scared of the paper industry still."[17] Indeed, the economist David Vail referred to the paper industry during the period as the "third house of the Maine legislature."[18] For their part, the initial reaction of the state's paper companies to the film appeared to be somewhat reserved. "So far, except for stories in the *Bangor Daily News*, the film hasn't stirred up much public interest," read an entry in the newsletter of Great Northern Paper Company shortly after its release. "Where showings are held in Northern Maine, we plan to send representatives to balance the facts," further accentuating distrust between loggers and the mills.[19] In the following weeks, as *Cut and Run* received more attention from the state's citizens and media, Great Northern's newsletter took greater notice, stating that, after a showing of the film to 150 in Fort Kent, "the program included spirited panel discussion in which foresters and contractors branded 'Cut and Run' as a one-sided approach to the issue," with one conservative news outlet apparently referring to it as "a hard core socialist propaganda film," GNP apparently taking a shot at the film's underlying pro-labor tone.[20] As attention to *Cut and Run* continued to pick up steam, a couple of months later Great Northern's newsletter read: "'Cut and Run,' the movie which has raised the ire of the forest industry . . . is critical of the life of today's loggers, blaming all of the problems on paper companies," further demonstrating the documentary's more lasting impact on the public image of Maine's forestry sector.[21]

In one segment of the film, a group of malcontent and apparently unemployed Maine loggers is interviewed. A spokesman for the group refers to immigrant and bonded woods workers from Québec as "hassle-free employees" because they worked long hours and did not demand employee benefits. He goes on to claim that Québécois immigrant loggers work for less money and are assigned the best logging opportunities, or "chances," asserting that "it's common knowledge from here to Fort Kent that Americans always get cleanup, Canadians always get the big wood. And any lumber camp you go to from here to Fort Kent and south as far as Patten, it's the same story."[22]

However, perhaps inadvertently supporting the notion that Québécois loggers were more productive than their American counterparts, the group's spokesperson states that "in a small situation like this you have four Canadian crews cutting roughly sixty cords a week, where we may have seven American crews and we don't even cut close to that." It is unclear from the interview whether the relatively meager productivity of the American crew was due to poor work habits or to their being assigned "cleanup." Nevertheless, the rather dawdling demeanor of several of the complaining American loggers suggests that poor logging chances may not have been the only reason for their lack of productivity, despite comments from an MWA activist some thirty years after the strike praising the group's leader. "He's the really articulate one that says [in *Cut and Run* that] 'they're gonna shut the woods down before they ever let us organize,'" she said, referring to the mills and industrial landowners.

A transplant from Pennsylvania, the group's spokesperson gave up on logging and "went on to sell cars," according to the MWA activist.[23]

The documentary's narrative estimated that, by the late 1970s, approximately 1,700 of Maine's estimated 7,000 woodsmen were employed directly by the paper companies, with the rest being independent logging contractors and their employees, because "by and large, it was much cheaper for (the paper companies) to receive wood from the independent woodsmen than to cut it themselves." *Cut and Run* described the work habits of Québécois woods workers this way: "At present, about 2,500 Canadians leave their homes late Sunday or early Monday. They work in the woods until Thursday afternoon or Friday morning and then return to Canada. They cut about one-third of the pulp and timber harvested in Maine. A majority of these men work in Maine on commuting visas, but there are still a great number who are bonded to a specific employer on a yearly basis. These men are called 'bonds.'"[24] Another view of the tensions around the activity of the Maine Woodsmen's Association in the mid-1970s was offered in the 1977 *Report on Bonded Canadian Labor in New England's Logging Industry*. Referencing the MWA's October 1975 work stoppage, the report focused on the two questions raised by the U.S. Senate's Select Committee on Small Business, which held hearings in Presque Isle, Maine, on the matter: Why is there a shortage of woods workers when there is a high unemployment rate? And why aren't there qualified and willing workers for the woods? According to the report, "The MWA contended that employment of Canadians had depressed wages and working conditions and made it more difficult for domestic woods workers to obtain full-time, year-round employment," an argument that has been echoed several times since that time. Defining bonded woodsmen as "Canadian nationals who are permitted to enter the United States at the solicitation of US employers following a procedure which certifies that no domestic labor is available to perform the type of work for which they are being sought," the report outlined the criteria for certification, including that "the employment of foreign labor will not adversely affect wages or working conditions of domestic workers."[25]

From its perspective, Great Northern Paper company summed up the situation in its October 17, 1975, newsletter this way: "On October 5 pickets first appeared at the gates of paper and lumber mills. The story dominated the television news reports of the evening. The [*Bangor*] *Daily*

News on October 7 headlined the story "Woodsmen block trucks to three plants." Picketing members of the [Maine Woodsmen's] Association told newsmen they wanted more money for the wood they cut and hauled to mills." The newsletter outlined the spread of the strike across several of the state's paper mills, with Great Northern becoming "the center of attention from the news media when company representatives met with [MWA president] Mr. Birmingham and members of his Executive Committee in Millinocket."[26] A month later, Great Northern summarized its side of the story in the aftermath of the MWA's strike: "During a U.S. Senate hearing in Presque Isle on November 8, a GNP representative stated "Until we can develop a stable labor force of Americans willing to harvest the wood which has got to be cut in remote areas, Canadians are the only people available to cut substantial amounts of wood needed by Great Northern and other companies . . . Without the Canadian timber harvesters, the forest industries would face a staggering wood shortage,"" a common refrain among the state's paper companies.[27]

What exactly the MWA was hoping to accomplish was a subject of some controversy. When asked by a legislative committee what it wanted to achieve, the organization's president, Wayne Birmingham, put it like this: "All we want the Department of Labor to do is to enforce the law as it is written." That is, members wanted to "require the employers to make a good faith effort to hire Americans, require them to pay the rates that are the minimum rates under the regulations and that are supposed to be in the contract, and require them to give us the contract," with the employer providing the equipment needed to do the job, as per the law.[28] According to the MWA's legal counsel, "While we were reluctant to be, you know, anti-Canadian, we saw it as a wage problem . . . By and large, the leadership of the Maine Woodsmen's Association was interested in higher wages for them and everybody, not simply the exclusion of fellow workers." As attorneys, "we were trying to show that the employers were manipulating the visa system in order to get as many Canadians as they could in place, and that that was their first choice, not their last choice."[29]

Nevertheless, some in the MWA were more militantly anti-Canadian, while others were focused solely on better rates for delivered wood, rather than on correcting alleged violations of labor laws in the state's north woods by the paper companies. This appeared to be the case initially in southern Maine, where there were fewer Canadians in the woods than in the north. Southern Maine loggers organized themselves and rallied around improving the price of their wood, rather than the enforcement of labor laws that were already on the books and that would reverse paper company preferences for hiring Canadians—laws that they saw as having little or no relevance to them. Although some MWA organizers asserted years later that "this was not an anti-Canadian labor issue," others strongly maintained that it was.[30] This assertion was later confirmed by a founder and leader of the MWA, who allegedly stated during prestrike MWA meetings that "we gotta get rid of the Canadian labor." Regarding the October 1975 strike, he stated that "I think it was Birmingham's initiative that we go on strike . . . he wanted to do something, so we did." The strike was needed, but "we were arguing the wrong points . . . [We should have struck on the issue] of Canadian labor in the woods being used to displace [American loggers] and cutting practices [of the paper companies]. Neither of those was a topic,"[31] demonstrating the divide within the MWA about the motivations of the leadership and what the strike was truly trying to achieve.

Offering a different perspective, here's how the *New York Times* of October 19, 1975, covered the "bitter warfare" between Maine's loggers and the paper industry as the strike was unfolding:

> For two weeks, members of the Maine Woodsmen's Association have joined in a statewide work stoppage aimed at winning higher earnings by forcing the mills to recognize the association as the bargaining agent for woodcutters and pulpwood truckers . . . The issues underlying the strike include the use of Canadian labor in the woods, state restrictions on maximum truck weights and the rising cost of everything from skidders to saw blades. At the heart of the controversy, however, is the relationship between the pulpwood cutters and the mills. The woodsmen, referred to as "independents" because they are not employed by the paper companies, insist their independence exists only in name. They say that the mills have them in an economic stranglehold, dictating the quantity of wood they cut and the amount they are paid. The mills, on the other hand, look at the woodsmen as private contractors or small businessmen.

> Mill managers defended their refusal to bargain with the woodsmen's association by saying that such a move would violate Federal antitrust laws. A Maine judge upheld that position five years ago when he ruled that a woodsmen's boycott constituted a conspiracy in restraint of trade. That brief organization effort, much smaller than the one now underway, soon dissolved.[32]

According to the *New York Times* report, Birmingham claimed that the MWA's membership increased from about four hundred to as many as three thousand loggers once the association took to the picket lines on October 6 of that year. Birmingham's various estimates of the strength of the MWA have been questioned, however. The estimate of three thousand MWA members, for example, may have derived from the size of the MWA newsletter mailing list, rather than a count of the number of active MWA members during its civil action against the mills. Media estimates at various demonstrations and attendance at meetings pegged the MWA's active membership in the hundreds, not thousands.[33] Whatever its representation of Maine's logging workforce might have been, essentially the MWA "accused the mills of hiring just enough Canadians to hold down the price paid to the independents," the *New York Times* reported. However, the often divergent motivations of the "strikers" were more complicated than that.[34]

As the movement gained statewide momentum and more loggers joined in, "before we knew it there were five major paper companies that were being struck and couldn't get wood at the price they wanted it . . . It was a spontaneous objection and outcry from people who were tired of working at the low price." The five paper companies responded by bringing lawsuits against the MWA and its members for "conspiring in restraint of trade," including price fixing and violations of the Sherman Antitrust Law. Hearings took place in Bangor. The governor interrupted the hearing and in a rant against the MWA demanded an immediate end to the strike. According to the source, because only federal courts can enforce the Sherman Antitrust Law, the case was removed to a federal court. To avoid delays with federal court scheduling and proceedings, paper company attorneys withdrew their federal case and refiled it in state court, this time not mentioning the antitrust aspects of the case.

Legal counsel for the striking woodsworkers characterized Longley as "very backward and ignorant" in his rather bombastic interference in the case, caring "nothing about [the striking loggers'] legal rights, procedural or substantive."[35] In the end, MWA president Wayne Birmingham gave in to the governor.

In the meantime, at least in part owing to discrepancies and a lack of clarity in the MWA's objectives among its loosely organized membership, the strike was beginning to unravel, with some loggers wanting to go back to work amid reported acts of sabotage of paper company equipment. Indeed, the *New York Times* reported that loggers opposed to the strike went to Governor Longley complaining of death threats and destruction of their logging equipment. "The police reported an apparent attempt to blow up a section of railroad track serving a mill, and state troopers started to escort mill-bound pulpwood shipments. Several arrests were made," the report stated.[36] According to one MWA organizer, strikers "were being unruly . . . it wasn't a well-reasoned effort . . . I began to realize that there was very little we could hope to gain. We made a statement, we made a show, but they weren't going to give in to us."[37] Apparently, some striking workers had also become concerned about their futures. "I had 50 men on a picket up here. They were taking pictures of us and identifying every one of us. Hey, they essentially blackballed you. They may not have admitted it, but they did."[38] Adding to the challenge of maintaining some cohesion among the MWA's membership, loggers were isolated not just geographically, with workplaces "off the beaten track," but also by the nature of their work—working long hours each day and traveling long distances to their work sites, leaving little or no time for anything else. Making matters even more tense, paper workers were apparently up in arms about the strike because they were running out of raw material, jeopardizing their jobs in the short term. The governor intervened again, and the paper workers union put pressure on Birmingham. "Before you know it, the paper workers union (later corrected to 'the carpenters union') has hire Wayne Birmingham to be on staff to organize the wood cutters," and eventually the MWA labor action simply fizzled.[39] "Nobody feels whipped," Birmingham declared to the *Times*. "We've only begun to fight. This isn't the end. This is only the beginning." For perhaps the first time during the MWA fiasco, he may have been right.[40]

"THE WRONG ENEMY"

In the near term, the MWA work stoppage temporarily curtailed production at several wood-processing facilities in Maine, damaged track owned by the Bangor and Aroostook Railroad that carried pulpwood and wood chips to the paper mills and paper from the mills located along the Penobscot River, and blockaded the flow of wood at the Canadian border. According to David Vail, "The MWA's tactic was to embargo timber deliveries to the mills, a resolution backed by mass picketing at the mill gates. The strike (as they called the embargo) lasted three weeks and was a failure. The paper companies made no concessions on the central issues. MWA's strident rhetoric and violence against recalcitrant loggers eroded public sympathy. The members became divided and demoralized, and MWA soon disappeared as a major force in politics and the woods."[41] The collective action by the MWA was impeded not only by labor laws but also by logging's isolated workplaces spread across vast northern Maine forest landscapes; the increasingly independent nature of the logging enterprise; and a nominally mutualistic, but lopsided, relationship with wood-consuming mills. In addition, despite a strong culture of frontier autonomy and self-reliance among the region's logging community, Maine's north woods was also plagued by a pervasive sense of victimization. According to one MWA leader and its president in the years following the strike, "We didn't know how to negotiate and we didn't know anything about organizing, so they took real advantage of us. We got [an additional] $0.50 a cord and went back to work. Which was a ridiculous thing." In the end, "the Maine Woodsmen's Association lasted only six years; it lasted from 1975 till 1980 . . . It was useless . . . They had us under their thumb; they had beat us."[42] Commenting on the collective action taken by the MWA against the state's paper companies, one observer summed up the situation this way: "It was precisely the complexity of relationships in the timber procurement network that prohibited the MWA from acting as the agent for wood workers as a whole . . . [T]he paper corporations had promoted this complexity with the knowledge that such legal obstacles would ultimately prevent loggers from mounting an effective challenge to their power."[43] Whether the author's assertions reflected the on-the-ground reality leading up to the MWA action is unknown, since there appears to be little publicly available documentation of the perspectives of the paper companies.

Agitation and violence against Canadian woods workers by MWA members continued, however. In 1975, shortly after the failed strike, the *New York Times* ran a story describing a gathering of loggers in front of their logging camp in Maine's north woods. Lamenting the outcome of a recent union vote, "they spoke of their defeat in the soft, Scotch-Irish accent of people isolated for generations. 'Damn Canadians,' said one. 'Only one thing left,' said another, and there were looks and knowing laughter around the circle. 'Sabotage,' said another man softly."[44] Denials by some that the MWA and the strike were not, at least in part, related to an anti-Canadian sentiment among many who worked in Maine's logging woods were not holding up. That the MWA was originally formed, according to a 1977 report of the Massachusetts Agricultural Experiment Station, "to represent domestic woods workers"[45] and not all the state's "woodsmen"—including Canadians—further supports an exclusionary posture that's difficult to square with contentions by some of MWA's members that they had nothing against Canadians.[46] "What we were trying to do was not exclude the Canadians [from working in Maine's logging woods]," according to MWA's legal counsel. "We were trying to get the employers, if they wanted Canadians, to first offer the jobs to Americans as they were required to under these favorable conditions that were required by law and, by God, we'd take the jobs."[47]

The MWA, a loosely organized association to begin with, faded away several years after the strike, but apparently not before allegedly inciting further violence in the state's logging woods. One member described a raid by MWA loggers on First Lake Camp during the summer of 1978, almost three years after the strike: "A group of woodsmen got together from MWA people and went down there and basically told the Canadians to go home. They busted up some toilets and the door, too, in the shed to scare everybody . . . I didn't think it was the cleverest thing that could have happened." After an investigation into the incident and indictments that followed, the organization became preoccupied with raising legal fees instead of organizing. "It certainly hurt the perception of the organization; it took away a lot of momentum. I think it kind of was the end of the MWA as a labor organization, in my mind . . . That [incident] was just a real bad thing . . . for the organization,"[48] further amplifying anti-Canadian sentiment in Maine's north woods. According to one MWA organizer, lamenting the animosity toward Canadian loggers, "So many of the more Yankee woodsmen identified the Canadians as the enemy. What a waste of time, what a waste of precious time . . . I guess I would tell young people, 'Don't waste your time on the wrong enemy, whatever you do.'"[49]

At the time, the MWA strike appeared to have little impact on the logging community of the broader northern forest region. For example, in its November 1975 issue, the *Northern Logger*, the industry trade journal for the northeastern United States, offered only a couple of passing observations about the MWA's action against the mills, including that the strike was "punctuated by reports of violence and intimidation" and that the striking loggers' issues included the use of Canadian labor in the woods, Maine's restrictions on maximum truck weights, and the rising cost of logging-related equipment.[50] The trade journal did not publish a feature article on the MWA strike. In addition, the 22nd Northeastern Loggers' Congress, sponsored annually by the Northeastern Loggers Association (publishers of the *Northern Logger*) and held in April the following year in Glens Falls, New York, did not formally address the MWA strike that had occurred just six months earlier.

However, it was likely not a coincidence that, less than six months after the MWA strike, a representative of the U.S. Forest Service presented the following information on the distinctions between employees, employers, and independent contractors to professional foresters at the Annual Meeting of the New England Section of the Society of American Foresters on March 10, 1976, in Boston, Massachusetts: "Workers who are classified as independent contractors rather than employees may not form labor organizations for purposes of collective bargaining. They also may not picket and engage in boycotts in restraint of trade. And employers have no responsibility under the National Labor Relations Act toward such workers . . . The primary legal test that is used to answer these questions . . . is whether or not the right exists to control both the results of the work and the method of performing it." According to the presenter, other tests included the amount of personal investment by a worker in purchasing and maintaining equipment; the opportunity on the part of the worker for profit or loss; the amount of skill required to perform the work; and whether the worker is paid by the job versus time or piecework. In each circumstance, the more that a worker demonstrated these attributes, the more likely it would be for the worker to be considered an independent contractor by the legal system at the time, according to the Forest Service.[51] Given the timing of the presentation in the immediate aftermath of the MWA strike, it appeared that the Forest Service was reaffirming its position on the independent logging contractor phenomenon, offering few, if any, ideas on how to relieve unresolved tensions between loggers and mills.

What, if any, were the long-term outcomes of the MWA's labor action? According to one organizer, the strike "amounted to being a statement." The *Washington Post*, for example, "wrote national stories" about the strike, boasted an MWA organizer.[52] Unfortunately, the "statement" made by the MWA strike was not always one that would reflect well on or elicit much sympathy for the organization and its members. A *Washington Post* story published two years after the strike told of Maine woodsmen who began "to prowl the night, launching raids" against Canadian woods workers. Referring to the MWA as "militant," the article described an event that took place on the night of August 17, 1977, when "some 40 Americans arrived in pickup trucks at the logging camp of the St. Regis Paper Co. . . . and chased 19 Canadian loggers out of the gray shacks and trailers and into the woods. The Canadians were back the next day, unharmed but badly shaken. St. Regis . . . promptly filed suit against eight [of the Maine] woodsmen. It charged that intruders carrying axes, clubs, and bats 'terrorized' the occupants of the camp and threatened to burn down the buildings unless the Canadians fled."[53]

Others described the MWA strike as a wake-up call for the paper companies and that perhaps the most important outcome was in subsequent legislation related to what they saw as improvements in log scaling and forest management practices in the state's industrial forest. It may be argued, too, that the right of Maine's loggers to collectively bargain, established in state law in 2019, traced its origins, at least in part, to the efforts of the MWA some four and a half decades earlier.

In the aftermath of the strike and the ultimate demise of the MWA, several organizers reflected on the MWA's impact and the future of logging in Maine, especially in the state's north woods. According to a key Franco-American MWA activist, reflecting on woods workers she met while helping lay the groundwork for and organize the association , "I never met an atheist. They all were in awe of the forest . . . The human relationship to that land, to me, is extremely profound. I think that's something that we're losing and it's very hard to reverse that; it's not simple at all." Comparing woods workers with unionized workers in the paper mills, she continued, "That's the difference between the mill workers and the woodsmen. Woodsmen are not bitter or cynical. They're pissed off, they're kicking tires, but they're not like that bitter, cynical stuff. Funny, uh? I don't know why."[54]

A logger who was involved in the organization of the MWA tried to explain the loss of a sense of community in the logging woods and in the remote towns that loggers and their families helped to build and support: "One of the biggest losses is cultural. That a state that's 90 percent forest . . . there's hardly anyone to work in the woods anymore . . . That whole way of life is pretty well gone. There's still some people who make their living in the woods, but it's not the same thing as having towns whose life was based in the woods and families whose life revolved around the woods," the logger asserted. "So those towns are disappearing, or they're changing into something else, into places for vacation homes and tourists, which is very different . . . The trees grow back . . . but that way of life I don't think is going to come back."[55] These issues—logging capacity, changes in forestland use and forestland ownership, impacts of a rapidly changing forestry sector on rural communities and culture—would all continue to challenge an already declining industry. That Maine is 90 percent forested would soon mean little to a dysfunctional industry already in decline.

Predictably, the strike also spawned several studies and reports. However, although there were a number of postmortem assertions and allegations about the troubles in Maine's logging woods, very little was substantiated, serving only to further inflame and perpetuate argumentation about Canadians working in Maine's woods. Even a 1977 study "encouraged by" the regional office of the Employment Training Administration of the U.S Department of Labor, with the sweeping title *A Report on Bonded Canadian Labor in New England's Logging Industry*, relied heavily on anecdote to reach some of its conclusions. The report also cites "*The Paper Plantation*, a report based largely on the Bowdoin College, Public Affairs Research Center [PARC] study of labor for timber harvesting in Maine." Yet, there is little original or empirical research in the PARC report, which is mentioned only very briefly in the nearly three hundred pages of *The Paper Plantation* and which offers little more than opinion based on a historical synopsis of the situation up to its publication in 1974. Nevertheless, and despite clear shortcomings in the gathering and analysis of primary data, the 1977 report concluded, "based on subjective judgment from persons in firms that hire woods workers and individuals in the five regional employment offices of the Maine Department of Manpower Affairs who have responsibility for woods labor requests . . . there is not an adequate domestic labor supply to fulfill demand" for woods workers in Maine. Remarkably, the *Report on Bonded Canadian Labor* also referenced an article from *Yankee Magazine* that suggested that "a high proportion of the population in northern Maine are of French origin and speak French."[56]

The matter of the MWA strike's legality and its violation of U.S. antitrust laws was distilled a couple of years later after a collective action by the Michigan Independent Wood Producers' Association in 1977, primarily over the familiar contention that wood prices were too low: "Can an independent businessman maintain the advantages of his independence and at the same time gain the strong government supervised collective bargaining benefits that are by law available only to bona fide employees?" questioned one writer in 1979, four years after the MWA strike.[57] Although it took some time and further controversy and conflict to unravel, in Maine the answer to that question would eventually be "yes."

BROKEN BONDS

Tensions between Maine and Québécois loggers stubbornly persisted during the decades that followed. In October 1998, for example, more than a dozen northern Maine loggers used their pickup trucks to block several private roads at the Saint-Pamphile border crossing that were points of entry between Maine and Québec, used by Canadian logging trucks to access industrial forests in northern Maine. The weeklong blockade was organized "to protest the illegal hiring of Canadian loggers over hardworking Mainers," according to one of the Maine loggers involved, asserting that Canadians could work for less because of the prevailing exchange rate and their national health care system. The Maine loggers stepped aside when the state police got involved, and the U.S. Department of Labor agreed to investigate their complaints. According to the Maine logger, "by the time I was working in the woods, large landowners had already systematically stripped away workers' rights by making loggers and haulers independent contractors instead of employees. At the time, folks had no idea how it would threaten our livelihoods." Of the blockade, he added, "All we wanted was a chance to provide for our families, doing the work our fathers, grandfathers and great-grandfathers did before us. But we didn't stand a chance in the fight against corporate landowners."[58]

A year after the 1998 Saint-Pamphile protest, some loggers from northern Maine's Allagash region blockaded a border crossing between Québec and Maine. Echoing complaints voiced in 1975, the Allagash loggers claimed that Canadian loggers were taking American jobs and were depressing wages for local loggers because Canadians were willing to be paid less,[59] with the difference in the exchange rate between the two countries at the time cited as a contributing factor in the perceived inequities. In addition, some Allagash loggers complained that, at a cost of $5,000, the state's Sustainable Forestry Initiative–recognized logger certification program was prohibitively expensive.

Short on data to verify the contentions of northern Maine's logging community, the state began to look into the matter. A study commissioned by the Maine Department of Labor the same year concluded that although a shortage of workers existed in northern Maine, the bonded labor program did not depress logging wages statewide but likely had a "highly negative effect" on wages for some loggers in northern Maine.[60] Four years later, the Maine Legislature's Joint Standing Committee on Agriculture, Conservation and Forestry Logging Payment Study summarized several issues raised by loggers during a meeting at the University of Maine at Fort Kent on September 12, 2003, including challenges associated with securing financing for logging equipment; one-sided, non-negotiated contracts for logging services; and inequities in hiring practices related to the H-2 bonded labor program that kept logging wages low.[61]

The following year, the report of the Maine Legislative Committee to Study New Payment Models for the Logging Industry concluded that the "availability of Canadian loggers has had an adverse impact on Maine workers in the St. John Valley." Later that year, state legislators from northern Maine urged the state's congressional delegation to review the cap on bonded woods labor, stating that "there was a time for bonded labor, but reliance on foreign workers has created a system that keeps

Americans unemployed . . . and is responsible for depressing logging wages for Americans."[62] To drive home their point, on August 6, 2004, two Maine state representatives wrote an invitation to the Maine congressional delegation to attend a meeting of the International Loggers Association in the small northern Maine town of Eagle Lake about the bonded labor situation, in which they asserted that "bonding is supposed to supply workers when there is an actual shortage of labor. In reality, bonded labor is used to depress logger wages. Contractors advertise operator positions for equipment they do not own. Americans are turned away from these jobs because a Canadian worker comes with his or her own equipment and will do the work for much less than an American. Canadian workers are also covered under Canada's national health plan, making them cheaper workers to hire."[63]

Throughout the early 2000s, Maine's forest products sector lobbied the U.S. Department of Labor to allow hiring of additional Canadian bonds on H-2A visas, expanding guest worker status to more immigrant loggers than would be allowed under the H-2B program, since there is no annual cap on the hiring of H-2A visas. Some north woods loggers pushed back, claiming that expanding the H-2A program to include woods workers would disadvantage American loggers, as it would essentially give the state's logging contractors the ability to increase the number of bonded woods workers in Maine's logging woods. Nevertheless, in 2008 and again in 2010, the U.S. Department of Labor released a rule that included logging in its definition of agricultural labor and services. Among other federal regulations, American logging contractors hiring H-2A visa workers were required to show proof of ownership of at least one piece of equipment, not including hand tools, for every two bonds hired under the program.

The hiring of loggers on H-2A visas gained national attention. The U.S. logging industry supported the claims of Maine loggers who felt disadvantaged by the hiring of Canadian woods labor. The executive director of the Texas-based American Loggers Council—"The National Voice for Professional Loggers"—weighed in on the bonded worker matter in 2004, stating that "the perceived labor shortage of U.S. workers is not because there are not enough people to fill the jobs, it is because the logging rates corporations are willing to pay to the loggers will not allow the loggers to hire American workers," echoing the refrain of some northern Maine loggers. Referring to the support by Maine's congressional delegation for the U.S. labor secretary's initiative to grant Canadian loggers entry to the United States on H-2A visas, he went on to assert that "the perceived 'labor shortage of U.S. workers' is not because there are not enough people to fill the jobs, it is because the logging rates corporations are willing to pay to the loggers will not allow the loggers to hire American workers. Supporting this 'reclassification' of H-2B visas will only promote importing more labor from Canada and displacing U.S. logging jobs."[64]

The following year, the *Wall Street Journal* ran the following story about restricting the hiring of Canadian loggers to work in Maine's north woods:

> By law, every job a Canadian gets must be first advertised in Maine at the government-certified rate, which on average is $27,000. Few Americans apply. A 1999 study commissioned by the state found that Canadians don't depress pay, except in pockets of the far north. American loggers, it concluded, simply hate camping out in deep woods.
>
> "Certainly, there's some level of wage that would attract American workers," says Lloyd Irland, a forestry consultant who helped write the study. But that would necessitate raising pay not only for a few hundred Americans who would replace the Canadians, but for the American loggers and thousands of others doing related jobs.
>
> "The question is, could our industry survive at that wage level?" Mr. Irland says. "My short answer is, no. We're in a very cruel market. Several of our paper mills have been in and out of bankruptcy. A lot of people are struggling to keep our mills alive."[65]

Over the next few years, that struggle would prove to be a losing battle, despite inexplicably optimistic assessments of the state's forest economy from Maine's forest industry leadership.

Loggers from Québec responded to controversies surrounding their right to work in Maine's north woods. In the summer of 2005, eleven Québécois loggers met in Saint-Georges, Québec, to talk about their profession. They discussed a variety of logging-related subjects, including woods labor, respect for their profession, logging regulations, and working in Maine. Agreeing with the gist of the *Wall Street Journal* article, one logger from the Beauce region of Québec who participated in the discussion noted that "it's too far for them [Americans] to come up [toward the border]—so why not have loggers from Québec do the work." Another stated that "it's not about taking jobs away, but geographically it makes more sense for us to go there. It's closer to home for us—those guys don't want to stay away from home too much." Referring to the difficulties associated with crossing the border to work in Maine, a third claimed, "To cross the border here is like the inquisition," to which another chimed in humorously, "It's not so bad. The customs officer went to school with us!"[66]

In 2005, the Governor's Council on the Sustainability of the Forest Products Industry was charged with developing "recommendations to improve the competitiveness of Maine's forest products industry nationally and internationally." Recognizing a less-than-friendly business environment overall in the state, the council was directed to "examine the state's tax policy and regulatory framework" that may have been impeding capital investment. Among other recommendations related to the logging industry, the council suggested establishing "a transition plan to reduce, and potentially eliminate, the need for foreign H-2B workers." Acknowledging that it would take some time to develop and implement "a long-term plan to meet the logging infrastructure needs of Maine industry with domestic labor," the council suggested that, in the near term, "because a continued shortage of loggers could have a crippling effect on the forest products industry, the industry and state should continue efforts to relax the federal cap on H-2B workers during this transition period. This partnership," the council continued, "also needs to encourage the federal Department of Labor to revise its prevailing wage standards for H-2B visa workers, including the use of their own equipment."[67] The report also cited shifts in forestland ownership between 1994 and 2004, from industrial forest owners to large nonindustrial private forest (NIPF) owners, while smaller NIPF ownerships nearly doubled and conservation groups more than tripled their holdings during the ten-year period, likely signaling a change in the ways forests may be managed in Maine in the future.

Work stoppages in Maine's forest products sector continued, however, generally around rates paid for delivered wood that were considered by loggers and truckers to be too low. For example, truckers boycotted Irving Company in Portage Lake, asserting their independence by not wanting to unionize officially. "They pulled a boycott . . . against the Irving Woodlands," according to one source, "because Irving wasn't paying them a decent rate. In a sense they tried to force Irving to sit with them and bargain, and Irving took a position that "we don't have to recognize you because you're not an accredited bargaining agent,"" reminiscent of elements of the 1975 MWA strike years earlier. Irving responded to the work stoppage by buying their own fleet of trucks. Most of the striking truckers needed to work "so now they're working for Irving for $12.50 an hour driving Irving's equipment."[68]

J. D. Irving Company, which owned some 1.2 million acres of timberland in Maine, was later accused of avoiding collective bargaining with independent loggers who supplied their mills, claiming that a 2004 state law allowing independent loggers to bargain collectively affected only Irving's operations, compromising the company's ability to compete. Irving laid off eighty workers, followed by a threat to lay off some three hundred more if the law was not repealed. Irving's work halt was called "blackmail" by some Maine legislators.[69] The law in question permitted collective bargaining by independent logging contractors when the company has economic control over more than 400,000 acres in a single labor market area, thereby protecting loggers in markets where there is little or no competition for raw material and the mill can name its price for raw material. Irving referred to the law as punitive and unfair. The Maine attorney general's office appeared to agree that Irving did not necessarily dominate local and regional labor markets in the state. A spokesperson for Irving asserted that the company's investment in training for its employees and contractors resulted in "an average 29 percent increase in harvester earnings since the investment began in 2004." In June 2009, the state legislature passed a bill "scrapping the collective bargaining law."[70]

In the meantime, the matter of immigrant labor in Maine's logging woods was far from settled. According to a 2009 report, "state labor officials allege that Canadian firms working in Maine routinely engage in illegal practices to discourage Maine loggers from applying for jobs. In some cases, officials said, firms have no apparent intention of hiring U.S. workers for jobs in Maine. As a result, the state is asking the U.S.

department of Labor to conduct an audit of the federal H-2A waiver program that allows companies to hire temporary foreign laborers when US workers are not available." Among the alleged illegal practices: logging firms that posted pay rates that are lower than what is eventually paid to Canadian workers in an attempt to discourage Maine loggers from applying; logging companies that allowed Canadian contractors to use their own equipment; and charging fees to American loggers to enter land to apply for jobs and ignoring applications for work from Maine loggers. Maine's labor commissioner described the illegal practices as "pervasive and industry-wide" and "severely harming Maine's logging industry."[71] Yet not everyone agreed that the illegal hiring of Canadian workers was to blame for the lack of employment in the state's northern forest at the time. A representative of the region's Forest Resources Association suggested that the poor economy was responsible for high levels of unemployment among Maine's north woods loggers. "Markets for paper and lumber are very poor, so there isn't enough consumption of wood to employ all the loggers out there," he suggested.[72]

Nevertheless, in the summer of 2009, the U.S. Department of Labor sent letters to twenty-seven Maine logging firms asking them to provide information related to alleged labor violations, including those pertaining to allowing bonded workers. Some Maine loggers, although supporting a crackdown on the illegal hiring of foreign workers, questioned whether this was a meaningful step toward resolving the state's persistent bonded woods labor controversy. Threatening another in what was now a litany of border blockades by Maine woods workers over the years, one logger claimed that the combination of competition for jobs in Maine from Canadian workers, low wood prices, and lack of enforcement of labor laws had pushed him and others to the breaking point. "We have been putting it off for a month," the logger said of a potential blockade. "After a period of time here, we have to do something."[73] Several months later, the U.S. Department of Labor (DOL) suspended two logging companies in northern Maine from participating in the uncapped H-2A visa program, initiated by the DOL in 1952 to facilitate the hiring of temporary, seasonal farmworkers in places where there was an anticipated shortage of domestic hires. The DOL alleged that the companies had violated federal laws that required them to recruit American loggers before hiring foreign workers.[74] According to Maine's labor commissioner, this was the first time that federal labor officials had "cracked down" on logging businesses in Maine for violations of the program.

More discord around immigrant labor in Maine's logging woods followed, however. In 2010, the governor signed An Act to Protect Maine Workers, which allowed penalties against contractors and landowners who "knowingly violate federal bonded labor laws in hiring foreign workers through the H-2A Foreign Labor Certification program," according to a report in the *Bangor Daily News*.[75] The passing of the act followed a year when there were 180 Canadians employed as bonded workers at a time when the Maine Department of Labor listed 900 unemployed woods workers. The law also tightened up loopholes that encouraged some Maine logging contractors to use Canadian workers who brought their own logging equipment and made bonded workers ineligible for Maine unemployment benefits.[76] According to a spokesperson for Maine's logging community, "H2A Canadian workers were taking $500,000 a year out of Maine's unemployment," whereas "only $200,000 a year was coming in from the people they were working for."[77] A few months later, three logging companies from Fort Kent "acknowledged to the courts that they had violated a Maine law prohibiting foreign loggers from bringing their own heavy equipment into Maine," according to a the *Bangor Daily News*. "The cases against the three firms did not involve recruitment of workers," the report continued. "Instead, the firms admitted in court that they violated the law requiring logging companies to provide equipment to any foreign or bonded laborers."[78]

The crackdown and subsequent calls to levy heavier fines on offending logging companies were not supported by the state's forestry community, including the Maine Forest Products Council, further accentuating a chronic divide between the state's forest industry trade group and its north woods logging community. In 2011, some Maine state legislators also opposed tightening restrictions related to immigrant woods labor. One legislator resurrected the claim that "employers favored Canadian labor in the woods because, frankly, they're more productive. They show up for work. It is closer to their home. They work harder. They complain less," echoing persistent sentiments expressed by others. Providing a different perspective on the matter, another legislator asserted that "the reality is bonded labor keeps the price of wood down, it keeps mills open and it keeps a market for people like my dad who sell wood."[79]

According to a media report, the forest and logging industry employed around two thousand workers in 2020, with logging companies operating at less than 50 percent capacity as the pandemic stalled the economy.[80] Still, the number of bonded workers in Maine during fiscal year 2019 was relatively small, with the federal government granting only thirty-nine guest worker visas in Maine to seventeen different companies, compared with hundreds per year in the early 2000s. Further demonstrating the lack of harmony among players in the state's forest economy, the Professional Logging Contractors of Maine (according to its website, "the voice of Maine's logging industry") and the Maine Forest Products Council (according to its website, "the voice of Maine's forest economy") did not support pushback from some of Maine's north woods loggers on the H-2 visa matter, contending that bonded woods workers were needed and that, during a period of considerable crisis in the forestry sector, the state's forest industry had larger issues to address. Despite significant improvements in oversight and enforcement of the bonded labor program and tensions around more pressing issues confronting the forest products industry statewide, the conflict over Canadian loggers working in Maine's north woods continued to be a distraction for an industry already in free fall.

BARGAINING WITH INDEPENDENCE

In its 2009 "Study of the Statewide Market for Forest Products Harvesting and Hauling Services," the state of Maine recognized three "general populations of harvesting and hauling contractors":

> Very large, diversified contractors who often own land, as well as harvesting equipment and trucks. These firms typically have many employees, they enter into large wood contracts with mills and they have generally been able to maintain relative financial success;
>
> Medium-sized contractors with large equipment "systems" who may subcontract portions of their operations to smaller contractors, as well as employing workers to perform the services; and
>
> Small "independents" who typically own and operate their own harvesting or hauling equipment. These owner-operators

> may work directly with landowners (e.g., roadbuilding) or under a subcontract with one or more larger contractors.[81]

The report continued: "Harvesters in many areas of the state are now predominantly self-employed, generally getting paid by landowners strictly on a contract basis. We are told by some that the shift away from company crews and logging camps was largely complete in the 1980s, and the contracting for harvesting and hauling services has long been the norm. Stumpage sales are no longer the norm except in the southern part of the state," where nonindustrial forest owners are more apt to sell stumpage than engage in service contracts. Wood harvested under a service contract versus that through a stumpage sale was distinguished as follows: "Contract harvesting is strictly about performing a service. The landowner retains ownership of the harvested wood and sells it directly to mills and other outlets. Harvesters no longer share in the rewards when wood value is high; on the other hand, when wood values plummet, they are also not susceptible to the downside risk," failing to mention that when wood values are low, loggers are often forced to scale back or stop working completely, since landowners are less likely to sell stumpage or engage with loggers through service contracts until wood prices are more favorable. Equipment payments, however, continued.[82]

Regarding the phenomenon of wood-consuming mills divesting themselves of forestland bases and logging woods employees, the report noted that "integration at the landowner to mill level is now almost gone in Maine. In the past a single entity generally owned the land, mills and labor force. Today, due to extensive industry land sales, this type of vertical integration is the exception and no longer the rule. Landowners are purchasing labor and selling to mills."[83] Doing so enabled mills to concentrate resources on manufacturing, separating themselves from the costs associated with forest ownership and management, while still competing on the open market for stumpage and harvested wood. The often-speculative nature of logging—including the vagaries of weather, stumpage availability, wood prices, and evolving environmental regulations, as well as the high cost of logging equipment—has resulted in serious challenges to logging businesses, especially those that are highly capitalized. "The principal barriers to entry for harvesters and haulers today are the availability of capital to finance equipment and the ability to obtain contracts to cut or haul wood . . . The heavy financial commitments involved in harvesting and hauling produce another effect—they increase the difficulty of exit from the business. Owners cannot afford to park equipment with a $5,000 per month bank payment when they deem rates offered for its services inadequate."[84] Anyone who has driven through the northern forest region has passed by dooryards littered with rusting skidders, feller-bunchers, and whole-tree chippers—the vacant storefronts of failed logging businesses—unaware of and likely reasonably ambivalent about whether the parked logging equipment is the result of a poor business decision by the logging contractor, broken promises from a wood-consuming mill or industrial landowner, wood markets that had abruptly and unpredictably tanked, or logging iron that had simply come to the end of its useful life.

Given the challenges associated with running a logging business, the problem of maintaining a workforce willing and able to work in Maine's north woods continued. In a 2009 interview with the *Bangor Daily News*, one forest industry representative reasserted the now familiar argument that "companies are having more difficulty finding Maine loggers willing to work in remote locations where commuting every day is not an option . . . To go through all of the scrutiny of bringing bonded workers in, I think it would be hard to discriminate against an American worker."[85] The following year, An Act to Protect Maine Workers, sponsored by a state senator from northern Maine who was a log hauler, passed into law in April. It required the Maine Department of Labor to be more active and vigilant in screening the bonded labor situation. The *Bangor Daily News* covered it this way: "Under the new law, any employer seeking to hire bonded workers now must go through a centralized employee 'clearinghouse' that will evaluate and refer potential logging workers . . . The law also requires the Department of Labor to maintain a list of contractors who are seeking to hire bonded workers and the landowners to hire contractors from that list."[86]

Fifteen years after the publication of a 2004 study by forest scientists from the University of Maine titled "Who Will Log," which cautioned about shortfalls in logging labor in the northern forest region,[87] the problem of labor shortages in the forest products industry began to gather more traction. The executive director of the Professional Logging Contractors of Maine stated in 2019 that "the heart of the problem is profit margins for logging contractors," which had dwindled owing to the realities associated with running a highly capitalized business in a climate of market uncertainty, low regional unemployment, stagnant rates for wood harvesting and trucking, and relatively weak forest worker wages.[88] According to the report, wages for logging equipment operators ranked lowest among those of comparable production-oriented occupations, and woods worker earnings had generally been stagnant over the previous decade.

In partial response to the logging labor crisis, and following decades of acrimony between loggers and industrial landowners over pay rates and bonded labor issues, An Act to Expand Application of the Maine Agricultural Marketing and Bargaining Act of 1973 to Harvesters and Haulers of Forest Products was passed by the Maine legislature and signed by the governor in 2019. As it already had done with Maine's farmers and lobstermen, the law allowed the state's loggers and wood haulers to bargain collectively with forest products companies and forestland owners who control both stumpage and wood markets. "This is about justice for Maine loggers and wood haulers, who have been used and abused for generations," according to the state legislator who sponsored the bill. "For too long, loggers and wood haulers have had no ability to negotiate, stand up for themselves and demand better conditions."[89] However, the impact of the act on the state's forest products industry remains uncertain. Given a forestry sector already in crisis, an industry that would continue to struggle if raw material costs increased, one representative from Maine's industrial forest asserted before the passage of the bill that "we all need to work together during these tough times as a united industry and create more opportunities for the state as a whole. This bill would set the industry backwards."[90]

But "working together" across its supply chain has not always been a strength of Maine's forestry sector. An attempt to pass similar legislation in 2004 also met resistance from industrial forestland owners and forest products manufacturers who claimed that it would increase costs for the industry. According to one northern Maine forest industry representative at that time: "We take the threat of this bill very seriously. It is a substantial threat to our success,"[91] further demonstrating the historical rift between Maine's loggers and the rest of the state's forest products industry on the matter. Signaling their intention to take advantage of their newfound right to bargain collectively, in 2019 a group of more than two hundred Maine loggers established the New England Loggers Cooperative.[92]

Given the forest industry's chronic argumentation around immigrant woods labor, it seemed likely that the evolving bonded woods labor and independent logger controversies in Maine's north woods would stubbornly persist. In the meantime, the state's forest products industry continued to operate in a relationship of nominal mutualism among loggers, wood-consuming mills, and industrial forest landowners, punctuated by episodes of dysfunctional antagonism that distracted the industry as a whole from achieving meaningful long-term success.

It all came to a head in 1975. Some of Maine's loggers struck. Some organizers of the movement—the Maine Woodsmen's Association—faced off against the wrong enemy, flashing a nativism that had been haunting Maine's logging woods for generations. Others took on the mills more directly, citing low pay rates for delivered wood and a lack of respect for their indispensable role in the forest products supply chain. Logging contractors bargained with their own independence, frustrated by their inability to make it work. Mills bargained with the long-term future of an industry and the workers and communities that supported it at risk, mustering their considerable wealth and political influence, bringing lawsuits against the MWA and its members for price fixing and violations of federal labor laws. The governor, described as "backward and ignorant," bullied his way into the fracas, siding with the mills and demanding an immediate end to the strike.

Work stoppages and blockades, logging camp raids, and busted toilets followed. Loggers' hard-fought independence, one of the things that they valued most about working in Maine's logging woods, at the same time deprived them of the benefits of being mill employees. A few loggers grumbled that all they wanted was to work as their "fathers, grandfathers and great-grandfathers did before us," as if a state's society and economy somehow owed them the privilege. For their part, some mills and wood procurement organizations continued to play one logging contractor against another, taking advantage of their status as independent contractors by encouraging an overbuilt logging workforce of competing wood suppliers. Others reportedly showed preferences to immigrant woods labor, further stoking flames of resentment toward Canadians.

It all came to a head in 1975. The mills may have won a battle, but Maine's forest economy was the real loser, its recovery far from certain. An industry in denial, and the workers, families, and rural communities that depended on it, continued to unravel.

Part IV

Haywire

Madawaska paper mill lays off 100 employees

—*Bangor Daily News,* May 1, 2009

Maine paper mills fight to survive

—*Bangor Daily News,* July 10, 2009

Madawaska paper mill seeks tax abatement that could mean millions less for town

—*Bangor Daily News,* July 28, 2012

School budget, shuttered mill prompt jump in Ashland tax rate

—*Bangor Daily News,* August 26, 2013

Maine's forest and forest products industries are thriving and remain an essential part of our state's economy.

—Maine forest products industry spokesperson, "Maine's Forest Economy: A New Report," *MaineTree,* 2013

Bucksport mill closure hits home as families face uncertainties

—*Bangor Daily News,* November 3, 2014

Lincoln mill to shut down tissue machine, lay off 20–25 workers

—*Bangor Daily News*, August 31, 2015

Maine's logging industry confronts pulpwood crisis

—*Bangor Daily News*, December 31, 2015

We are pretty optimistic. It's going to be a new world. It's not going to be your grandfather's paper industry; it's going to be more like your granddaughter's industry.

—Maine forest products industry spokesperson, "Maine's Logging Industry Confronts Pulpwood Crisis," *Bangor Daily News*, December 31, 2015

Report: Maine's forest economy fell $1.3B since 2014

—Mainebiz.biz, September 20, 2016

Loggers suffer one-two punch with mill and biomass plant closures

—Mainebiz.biz, April 4, 2016

Madison paper mill shuts down

—*Bangor Daily News*, May 25, 2016

The forest products industry is reinventing itself.

—Maine forest industry spokesperson, "Maine's Forest Economy," *MaineTree*, 2016

Maine pulp and paper trade group dissolves after half a century

—*Bangor Daily News*, January 17, 2017

Aroostook County's last biomass plant will shut down in April

—*Bangor Daily News*, February 27, 2019

Sappi to cut 75 jobs at Westbrook paper mill

—*Bangor Daily News*, July 9, 2020

The Maine forest economy remains resilient and a strong economic engine.

—Maine Forest Products Council, 2021

Chapter Six
Who Will Log?

Our grandfathers and fathers were lumber men, and we feel strongly inclined to follow in their footsteps.

—G. W. Drisko of Jonesboro, letter, 1850, Maine; reported in Richard Wood, *A History of Lumbering in Maine,* 1935

It's dangerous, it's hard to make a living and everybody hates you? Who's signing up for that?

—Michael Snyder, Commissioner of the Vermont Department of Forest, Parks and Recreation, *Adirondack Daily Enterprise,* January 28, 2016

Workers needed: Low pay, no benefits, dangerous working conditions, performing work that has little or no occupational prestige, but a high degree of uncertainty and instability. Enjoy up to three months or more of mandatory unpaid leave per year. No Paul Bunyan wanna-bes.

—"A Dying Breed," *Maine Sunday Telegram,* January 18, 2004

Maine's logging community continued to face serious challenges related to its relationships with wood-consuming mills and the state's industrial forestland owners, yet other dynamics were in play that threatened the state's logging businesses.[1] These included shifts in forestland ownership and land use, urbanization and evolving attitudes toward forests and logging, misrepresentations of the forestry sector's economic realities, and persistent cycles of instability in markets for raw material, especially low-grade trees. What this all means for the future of logging in Maine is a matter of some speculation. But given the downward trajectory of the state's forest products economy, the outlook remains tenuous, despite expressions of inexplicable optimism from leaders of the state's forest products industry.

One of the most important issues facing Maine's logging community and, by extension, its forest products economy has been the erosion of the logging workforce, with little hope in sight and no cohesive, industry-wide plan to meet the real issues head on. Some have pinned the problem on logging's lack of occupational prestige, too often ignoring the fundamentals underlying its lowly status—low pay, poor employment benefits, and lack of year-round work. Efforts to resuscitate the state's forest products industry have generally fallen flat, with taxpayer dollars subsidizing logger training for jobs that few really wanted. While contrived solutions, such as providing more logger training, may have improved on-the-job safety and productivity, the question of who will log in Maine's woods has stubbornly persisted.

THE BREAKDOWN

The matter of a shortage of loggers is clearly not a new one. In 1968 an extension forester from the University of Maine wrote in the *Northern Logger* that "the old days of logging are gone, and people don't know it. The old days were typified by men who would put up with seasonal work, low pay, rugged conditions, and old-fashioned personnel policies . . . My point is, a large segment of the population does not consider employment by the forest products industry to be desirable." Searching for solutions to the shortage, he suggested cleaning up logging's image: "First, we should be improving the image of the industry by getting up-to-date information to school teachers, service clubs, and to the general public by newspaper, radio, television, and any way we can do it." However, the author also recognized that this was not simply an issue of an industry suffering from a poor public image. He understood that logging as a profession—a career choice—had to change: "The woods worker of tomorrow must be provided with more regular work, fringe benefits, and better work conditions . . . If the forest industries are unable to pay an adequate wage and provide reasonably good conditions for workers and contractors, they cannot expect to have labor looking for employment in logging."[2] Nevertheless, Maine's forest industry continued to try to find solutions to its labor challenges that seemed to ignore widely recognized deterrents to a more robust workforce. According to one Maine forest industry leader, writing for the *Northern*

Logger in an article titled, "Appears That 1968 Will Be a Good Year": "Our trial use of displaced Tibetans has been encouraging and offers proof that properly motivated men can be trained to woods work."[3] On August 25 of the previous year, under the headline "Imported Workers Line Up for Jobs Americans Shun," the *New York Times* reported that "the Great Northern Paper Company, unable to find enough workers to bring down timber in Maine's rugged Aroostook County, found that Tibetan refugees, homeless and living in exile in India, were willing to come here." And headlines in Great Northern's newsletters proclaimed in July 1967 that "Tibetans arrive" and a year and a half later that a "Tibetan family joins Maine woods training."[4] In the half century that followed, there was no indication that Tibetan immigrants were making significant contributions to the state's logging woods labor pool.

So, where did Maine's loggers go? According to one observer of Maine's "woods labor problem" in the 1960s, the decline of agriculture in the state had contributed to the lack of people accustomed to challenging outdoor work. "Farm trained and farm occupied labor is virtually gone from the logging scene. There are many fewer farms than ten or twenty years ago and these farms are intensively managed. The typical farm is bigger now and the owner just cannot justify the time to work off the farm or even in his own woods."[5] About forty years later, a Massachusetts logger participating in a group discussion about woods work echoed that assertion: "The farms are gone, so the training grounds [for loggers] are gone."[6]

Commenting on the challenges of attracting workers to a profession that appeared to have little to offer beyond strenuous and dangerous work performed in an often-challenging forest environment, and at a time of transition to more mechanized logging operations, one observer wrote: "Manual woods work is among the most physically demanding imaginable. Couple this with the low prestige and low wages which manual woods work garnered, and it became more and more difficult to attract a reliable supply of committed workers as time went on and the long postwar boom created more attractive employment in other occupations." Offering ideas on how to address the challenge of logger recruitment, he continued, "To be reasonably attractive and workable, the occupation needed to offer secure, year-round employment; it required comparatively high levels of skill that would need to be offered in conjunction with attractive working conditions; and it needed to produce more wood at lower cost." Then anticipating the increasing mechanization of woods work, he suggested that "transforming woods workers into operators of sophisticated machines would address the problems of physical demands, the seasonal nature, and the low status of the work . . . Woods work would become a skilled and attractive occupation . . . A fully mechanized tree harvesting system would be characterized by higher productivity, lower costs, and higher wages."[7] The author's optimism about the benefits of mechanization in developing a stable logging workforce hasn't always panned out, however.

"CE N'EST PAS UN CHOIX"

Reacting to a reported shortage of loggers in the northern forest region, a 2004 University of Maine study looked at why loggers chose their profession and their sense of logging's public image. Using a survey with versions written in both English and French, the study reached out to all loggers who worked in Maine's logging woods, including over six hundred loggers from Québec who crossed the border to work in the state. Almost two-thirds of Maine loggers and three-quarters of Québécois loggers reported having relatives in the logging industry and said that they logged because they came from logging families.[8] Yet despite considerable intergenerational attachment to the profession, the authors found that most loggers who worked in Maine would not encourage their children to log. Writing about the breakdown of familial attachment to logging among Québécois loggers, one observer stated:

> *Les bûcherons désirent surtout que les jeunes profitent des meilleures chances qu'ils ont de s'instruire aujourd'hui et qu'ils acquièrent assez d'instruction pour ne pas devenir eux aussi des bûcherons et "passer par où on a passé" . . . En somme, les travailleurs forestiers désirent de plus en plus restreindre le nombre d'enfants . . . afin d'assurer un standard de vie normal à leur famille et un niveau d'éducation plus élevé à leurs enfants qui correspond aux exigences nouvelles du marché de travail . . . On désire avant tout leur éviter la misère qu'on a connue et*

> *qu'on continue de connaître. On ne veut pas qu'ils deviennent des bûcherons ou des femmes de bûcherons."*
>
> Loggers especially desire that their offspring take advantage of the best learning opportunities available and that they obtain enough education that they not also become loggers and "go through what they have gone through" . . . In short, forest workers want more and more to restrict the number of children they have . . . in order to ensure a normal standard of living for their family and a higher level of education for their children that is in line with the new demands of the job market . . . Above all, one desires them to avoid the misery that one has known and continues to know. One doesn't want that they become loggers or the wives of loggers.[9]

Further supporting the notion of a breakdown in intergenerational connections to the logging workforce, a Québécois logger who worked in Maine asserted in 2005, "There's less and less family in rural areas. Young people leave and so our history of family as the basis for forest work is changing." Suggesting that it's not simply a breakdown in familial attachment to logging but also a disruption in place attachment among young rural residents, he continued: "It changes the work because it's not so much about traditions anymore. It used to be that if you cut wood, you came from a rural area, and it's still like that a little bit, but you don't have to be attached to your place anymore."[10]

The marked erosion of familial attachment to Québec's logging profession was later supported by a 2008 Laval University study showing that a majority of loggers in the province would not encourage their children to become logging contractors.[11] Reasons given for breaking the intergenerational connection were uncertainty and instability caused by a steep downturn in the province's forest economy (40 percent) at the time of the study; difficult working conditions (23 percent); and financial challenges associated with running a logging enterprise (22 percent). Only 4 percent of Québécois loggers cited the negative image of logging as a contributing factor.

What about the retention of Maine's loggers? Did they intend to remain in their profession? According to one study, slightly more than half of Maine's loggers expected to stay in the logging business over the next five years, whereas approximately 24 percent indicated that they would not be in logging and 25 percent indicated that they were not sure. Among those respondents indicating intent to leave logging or uncertainty, reasons included "no money in it," "not profitable anymore," "market gets worse each year," "price of wood never goes up," " no wood and wood prices not keeping up with costs," "stumpage harder to find," "lack of benefits, insurance, and money," "getting too competitive and too mechanical," and "northern Maine may not need hand cutters in the future."[12] That logging may have been suffering from an image problem was rarely mentioned. The study concluded that logger recruitment and retention in Maine will have less to do with improving logging's image and more to do with better wages and employment benefits.

Given the challenges, why did loggers log? According to a study conducted in 2004, Maine resident loggers more often indicated that in choosing to become loggers, they felt less constrained by their education or the prospects of other employment than did loggers from Québec who worked in Maine, many of whom appeared to be more resigned to logging. In response to the statement, "I became a logger because I don't have the education for a different job," 26 percent of Maine loggers and 52 percent of Québec loggers agreed. Raising further questions about the future of the logging workforce in Québec and the implications for the future of immigrant labor in Maine's north woods, the study found that Québec resident loggers were older and less educated than their Maine counterparts. Indeed, when compared to loggers in the rest of the northern forest states, loggers from Québec reported over four years less education than loggers from New Hampshire, the most highly educated group. There also appeared to be more opportunity for alternative employment in Maine than in Québec. In response to the statement, "I became a logger because I could not find another job nearby," 19 percent of Maine loggers and 33 percent of Québec loggers agreed.[13] There were some similarities between the two cross-border communities of loggers. For example, a vast majority of loggers from both Maine and Québec indicated that they logged because they enjoyed working outdoors and the sense of independence associated with their work. However, fewer than half logged because it paid well.

Some Québécois loggers suggested that logging was the only work that was available or that they knew how to do. Put bluntly by one Québécois logger: "Ce n'est pas un choix. C'est le seul metier connu" (It's not a choice. It's the only job I know). However, most loggers in the region indicated that they logged because they enjoyed working outdoors and for its sense of accomplishment and independence, whereas others indicated that there were few alternative employment opportunities where they lived. Far fewer logged because it paid well, and a vast majority of Maine resident loggers worked without employment benefits.[14]

Responding, in part, to statewide concerns about logging labor and several University of Maine studies, in 2004 Maine's governor issued an executive order establishing an advisory council charged with several directives related to the state's forest products sector, including "identify[ing] strategies to support the workforce infrastructure needed to maintain a vibrant forest products industry . . . include[ing] an examination of the issues relating to the recruitment and retention of loggers, as well as other labor force needs." The *Final Report of the Committee to Study New Payment Models for the Logging Industry*, published in April 2004, focused on several study areas, including "current payment methods and new payment models for loggers." The report concluded that "availability of Canadian loggers has had an adverse impact on Maine workers in the St. John Valley" but that "recommended changes to the federal H-2 bonded labor program have not been implemented." The report went on to cite survey results "indicating that loggers are not encouraging their children to enter logging professions," suggesting "a wake-up call to the mills" and a greater recognition of loggers "as stakeholders in the forest products industry."[15]

A few years later, the state's Office of the Attorney General issued a report observing that fewer young people were entering the logging profession, in part because of the time and financial resources required to start up and manage a successful logging business.[16] Compounding the challenge, the report asserted, was the state's overall aging population and slow population growth. In 2012, a follow-up study of the logging community in Maine's counties that border Québec concluded that the fraying of intergenerational attachments among both American and Québécois loggers "may not only pose challenges for logging business stability and labor recruitment efforts in this region, but may also impact the economic vitality of the forest products industry as a whole."[17]

However, the challenge persisted, the governor's advisory council and state reports notwithstanding.

AN IMAGE PROBLEM?

As we discussed in chapters one and two, there was a great deal of early romance associated with logging and loggers, depicting loggers as not only prodigal and rambunctious but also daring, courageous, and carefree. Yet despite descriptions of larger-than-life loggers and the heroic nature of their work, logging continued to suffer from a poor image. One observer described the Canadian logger of the 1920s and 1930s as occupying "a rung quite close to the bottom of the social and economic ladder . . . he was considered to be a lumberjack for the simple reason that it was the only job he could get."[18] Another writer offered this description in the *Northeastern Logger*: "To the average city dweller logging is a colorful and romantic occupation. It deals with tall men, tall timber, and taller stories. A logger is depicted as a muscular giant who pits brawn and skill against the forest's mighty trees and nature's rugged terrain. While proud of the traditions, the modern logger is far from satisfied to remain a quaint character untouched by the benefits of a mechanical age."[19] That was in 1954. Later, loggers in Maine would be variously described as upwardly mobile petty capitalists and "rural sub-proletarians, academic failures and social misfits." Economist David Vail maintained that both Maine loggers and nonloggers alike hold timber harvesting in low esteem, asserting that loggers perceived a stigma attached to their profession among their neighbors.[20]

A University of Maine study attempted to sort out the logging workforce conundrum, noting that occupational prestige differs from occupational status, the latter referring to objective socioeconomic conditions associated with holding an occupation and the former describing subjective evaluations of the social standing of an occupation.[21] One source described the public's perception of the *prestige* of various occupations as being rooted in several relatively subjective attributes, including the degree of interesting and challenging work, intelligence required to perform primary work functions, and the intrinsic nature of the work.[22] Another suggested that nonmanual workers generally have higher prestige than do manual workers.[23] In contrast, occupational *status* is often measured by the more objective elements of education and income, frequently derived from census data.[24]

What did this mean for the logging profession? Not everyone agreed with the characterization of manual work as having low prestige. "Scholars should resist the temptation to equate strenuous manual labour with an absence of skill," wrote one observer in 1997. "Felling, for example, required enormous judgement to avoid hang ups and place logs in the desired position for skidding . . . Lacking any technological control of work, operators had to rely on the experience and 'working knowledge' of those they employed at every stage of the lumbering operation."[25] Nevertheless, despite historical descriptions of larger-than-life loggers and the heroic nature of their work, logging has long suffered from a poor image relative to other professions. For example, at the time of the Maine Woodsmen's Association strike, a study gave "logger" an occupational status score similar to "migrant worker," "packing house butcher," "unskilled garage worker," "unskilled factory laborer," "railway, airport porter," "laborer," "warehouse hand," and "road construction laborer."[26] Several years later, the category "lumberman, raftsmen, and woodchoppers" was given the same occupational status score as "newsboys" and "busboys."[27] A national ranking of occupations by their desirability, published in 2000, placed "logger" 248 out of 250 occupations, the prestige of an occupation linked primarily to whether the work performed was nonmanual (higher prestige) or manual (lower).[28] Contributing to logging's image were poorly cited historical threads denigrating loggers and their work that appeared to persist, despite data that would suggest a far different narrative about the profession and life and work in the logging woods of the northern forest.[29]

Similar challenges were faced north of the border. Challenging recruitment to Québec's logging workforce was an image of the Québécois logger that was not always consistent with the often highly technical reality of his work, according to one observer in 2005:

> *En fait, les travailleurs forestiers sont très majoritairement insatisfaits de leur occupation. Seulement trente pour cent des bûcherons se montrent satisfaits ou relativement satisfaits de leur occupation . . . Cette situation contraste avec le degré de satisfaction élevé (60%) observé dans l'ensemble de la population.*

> *Les bûcherons valorisent surtout d'autres occupations, telles que, par ordre de préférence, homme de métier ou cultivateur, puis ouvrier d'usine et employé de bureau. Il est à noter que toutes ces occupations permettent au travailleur de vivre avec sa famille . . . Selon la majorité, non seulement c'est le pire métier, mais "ce n'est pas un métier!" . . . Dans l'ensemble de la population régionale, cette main-d'œuvre est au bas de l'échelle socioéconomique. Elle se sent généralement, bien qu'à des degrés divers, isolée, ignorée, oubliée, rejetée et exploitée. Elle se perçoit marginalisée et surtout impuissante devant cette situation.*

> In fact, the forest workers are generally very dissatisfied with their occupation. Only 30 percent of loggers are very or moderately satisfied with their occupation . . . This situation contrasts with the higher level of satisfaction (60 percent) in the population as a whole. The logger especially values other occupations, such as, in order of preference, professional men or farmers, then factory worker and office employee. It is noted that all of these occupations allow the worker to live with his family . . . According to the majority, not only is [logging] the worst profession, it's not [even] a profession . . . In the regional population as a whole, this workforce is at the bottom of the socioeconomic ladder. It generally feels, although to different degrees, isolated, ignored, forgotten, rejected, and exploited. It sees itself as marginalized and especially powerless in the face of this situation.[30]

Interestingly, despite assertions about logging's poor image, the 2004 University of Maine study found that approximately two-thirds of Maine and Québécois loggers who worked in Maine's border counties with Québec became loggers because it was a respected profession in their communities, raising questions about the assertion that the region's labor problem could be explained solely by logging's poor image.[31] However, a vast majority of respondents from both sides of the border logged because they liked working outdoors and because they enjoyed the sense of independence. Consistent with reported lower levels of schooling among Québécois loggers, far more loggers from Québec indicated that they logged because they didn't have the education for a different job.

Familial attachment to logging was also stronger for Québécois loggers than for their American counterparts.

Studies of occupational choice in other regions of North America have demonstrated a strong sense of professional identity among loggers. For example, in 1995 a social scientist from Washington State described a sense of strong occupational self—the notion that an individual's self can be connected to a particular occupation—among the loggers he interviewed in the U.S. Pacific Northwest. He posited that loggers' occupational identity appeared to focus on four interrelated themes: independence, pride in skill, pride in facing danger, and a sense of being in a unique category of workers. Logging was also considered a source of identity for many woods workers. Occupational choice among loggers was based as much on the enjoyment associated with working in the woods and their familial attachment to the profession as on the income they derived from logging, similar to findings of the University of Maine study.[32] A year later, another social scientist from the region described the stigmatization of loggers as a relatively new phenomenon, citing Oregon as a state in which, in the mid-twentieth century, logging and loggers had been celebrated by its citizens in romantic and benign terms. She argued, however, that during the latter part of this last century, a sense of stigmatization has become pervasive among loggers in Oregon.[33]

But loggers' assessments of their profession, as well as the opinions of the general public toward logging, have been found to be quite variable across the northern forest region. For example, signaling a difference in how the profession is viewed among Maine versus Québécois loggers, almost two-thirds of Maine loggers, but less than half of Québécois loggers, indicated that the public views loggers as unskilled. In addition, although only 43 percent of Maine loggers agreed with the statement "the public respects loggers and the work that they do," over three-quarters of Québec loggers who worked in Maine agreed. At the same time, approximately two-thirds of both populations of loggers agreed that logging was a respected profession in their communities. "Who will log in Maine's north woods?" the authors asked rhetorically. "Due to its low social prestige, the logging workforce in the region likely will not be dominated by individuals who had not previously logged or who have little or no familial attachment to logging."[34] But did loggers and logging really suffer from low prestige among Maine's general public?

In the context of a labor shortage in Maine's logging woods, unsubstantiated assertions about logging's poor image beg the question as to what Maine's general public actually thought about loggers and logging. Describing occupational prestige as one's perception of social approval for a particular profession, a 2009 University of Maine study surveyed three hundred randomly selected Maine residents by phone to find out. Although most Maine citizens agreed that logging was a skilled profession and acknowledged its importance to a state that ranked first in the percentage of its economy being forest dependent, almost two-thirds would not recommend logging to a son or daughter. Respondents with less education, who were male, or who lived in northern Maine were more likely to recommend logging to their offspring. For example, 71 percent of female respondents would not encourage a son or daughter to become a logger, while 52 percent of males would.[35]

When participants were asked why they had not considered logging as a profession, 99 percent indicated that there were better jobs available in their area, while 91 percent cited work that was too physically demanding, 80 percent noted harsh working conditions, and 72 percent said poor job benefits. Others suggested that logging wasn't a healthy lifestyle and that it was too dangerous. Only 52 percent said that a lack of occupational prestige discouraged them from considering logging.

Importantly, additional findings suggested that a vast majority of Maine residents respected loggers and the work that they do; they understood the connection between wood products that they used and the work that loggers perform; and they recognized that logging can be good for the forest and is important to the state's economy. Respondents were more evenly split on logging's social prestige—48 percent of males and 54 percent of females said that logging has little social prestige. The study concluded that there were differences between loggers' sense of their occupation's social prestige and the opinions of the general public. For example, previous research indicated that Maine's loggers most often cited improving pay to attract new workers, whereas Maine's general public more often cited the nature of the work—for example, it being too physically demanding, harsh working conditions—for not considering logging as a career. Indeed, only 42 percent of Maine's citizens cited poor pay. The authors closed with this assertion: "Although the results of our study do not support a broadly held sense of low occupational prestige among Maine's general public, they do suggest a perception of often challenging working conditions and physical work functions required of loggers and a recognition of the skills required to succeed in a difficult work environment." In short, while a majority of Maine's general public did not appear to hold loggers and logging in low esteem, many just didn't consider it an attractive career path.[36]

NOT ONE MAINE

While there are clear differences between Maine-resident and Québécois loggers who worked in Maine related to their occupational choice and sense of professional prestige, what happens when we look at Maine loggers at a lower altitude? Are the views of loggers from the state's north woods the same as or different from those from Maine's more southern forests? The authors of a 2002 University of Maine study suggested that most previous studies of Maine's logging community had focused on the loggers of northern Maine, leaving questions about whether the loggers of the state's southern counties represent a distinct occupational community. The study compared the sense of occupational choice ("I became a logger because . . .") and prestige ("Describe the general public's perception of logging") among loggers who work in Maine's northern forest counties (Aroostook, Franklin, Hancock, Oxford, Penobscot, Piscataquis, Somerset, and Washington Counties) and the state's southern counties (Androscoggin, Cumberland, Knox, Kennebec, Lincoln, Sagadahoc, Waldo, and York Counties) (see table 2).[37]

Although most Maine loggers engaged in logging because they enjoyed working outdoors and the sense of independence and accomplishment, significant differences existed between the two populations on other matters related to their work. For example, there was a greater sense of resignation to logging as an occupation ("I could not find another job nearby"; "it was the best paying job available"; "I don't have the education for a different job"; "I come from a logging family") among the state's northern forest loggers than among their southern Maine counterparts, signaling caution about regarding the state's "logging community" as a single, homogenous population. The authors concluded that "implications for the future may include a continued decline in the availability of individuals who are willing to participate in a skilled domestic logging workforce, particularly in Maine's northern forest. This may be exacerbated by our finding that 72 percent of Maine's northern forest loggers would not encourage a son or daughter to pursue logging—in a region of the state that our data show has accounted for significant familial ties to logging vs. Maine's southern counties."[38] Indeed, while almost two-thirds of northern forest loggers agreed that they logged because *they came from a logging family*, only 45 percent of southern Maine loggers agreed with this statement.

TABLE 2. Comparison between Maine's northern and southern county loggers' reactions to statements listed as possible reasons for becoming a logger, as well as their sense of the public's perception of logging (N = 571 respondents)

	Northern Maine (%)	Southern Maine (%)
I became a logger because:		
I could not find another job nearby.	21	9
I come from a logging family.	65	45
It was the best-paying job available.	45	28
I don't have the education for a different job.	28	16
I have always done this work.	70	62
I enjoy the sense of independence.	94	97
I like the work.	98	99
I enjoy using skills to accomplish a task.	93	94
It's a healthy lifestyle.	71	77
It's challenging.	92	92
It pays well.	41	36
It gives me a feeling of accomplishment.	88	93
I enjoy working outdoors.	99	99
It is a respected profession in my community.	62	55
Describe the general public's perception of logging.		
The public views loggers as unskilled.	63	65
The public respects loggers and the work they do.	43	34

SOURCE: Deryth Taggart and Andrew Egan, "Occupational Choice among Loggers in Maine's Northern Forest and Southern Counties," *Adirondack Journal of Environmental Studies* 9, no. 2 (Fall/Winter 2002): 34.

Maine's general public was also divided geographically on whether they would encourage a son or daughter to become loggers, with almost half of respondents from northern Maine and only 35 percent of southern Maine respondents agreeing, based on the 2009 phone survey of a random sample of three hundred Maine residents. Residents of southern Maine also indicated that "logging has little social prestige" more often than their northern counterparts.[39] However, although a large majority of Maine's general public agreed with statements such as "logging is a skilled profession," "I understand the connection between wood products that I use and the work that loggers perform," and "logging can be good for the forest," respondents from northern Maine more often agreed with these statements.[40] "Although there are differences between people living in the more populated, less rural southern counties versus those from the rural, sparsely populated Northern Forest counties regarding whether respondents would recommend logging to a son or daughter, there were fewer differences between the two populations on their perceptions of logging and the logging profession," the study's authors concluded. Despite evidence of the general acceptance of logging among much of the state's population, the challenge of recruiting new workers to a sector suffering from a fracturing of its traditional intergenerational attachment to logging and a realization of better paying and safer work closer to home remained.[41]

Results of these studies had implications for employment and rural community stability related to forest products extraction and manufacturing in northern Maine at a time when employment in outdoor recreation and tourism appeared to be increasing, the authors suggested. Importantly, how would the state's logging community, and the forestry sector that depended on it, address the fundamental question, *Who will log?*[42]

TRAIN THEM AND THEY WILL COME?

Training has often been the assumption among those attempting to solve the logging workforce riddle, despite numerous studies and failed logger training initiatives that might suggest otherwise. As an answer to the question *Who will log?*, industry-sponsored logger training programs focusing on the recruitment of new loggers have had a history of limited success. According to Leo Thibodeau, employment manager for Great Northern Paper Company in Bangor, in his 1965 *Northern Logger and Timber Processor* article "Effective Recruiting and Training of Woods Workers and Loggers": "We felt that mechanization alone would not solve our long-range labor problems. Consequently, in the spring of 1950 we decided on a woodsman's training program. It was our hope that such a program would not only increase our dwindling supply of domestic woodsmen, but by proper indoctrination and intelligent training, produce a better and safer class of workers. We advertised extensively . . . Three [students] showed up. They stayed three days. In 1951 we tried again with the same results."[43] Four years later Great Northern Paper Company initiated another attempt to recruit woods workers through logger training, this time inviting "ten of the best qualified applicants as to age and physical fitness" to participate in a more formalized logger training program. The result: "Thus far, these experiments in training had cost us considerable money, much time and effort with not one solitary woodsman added to our labor pool. We had to conclude that many young men of this generation, regardless of social background, were definitely not interested in the strenuous work that is a part of the timberlands harvest," stated Thibodeau.[44] The industry was not discouraged, however. Consider this mid-1960s effort in Maine to build a logging workforce by introducing another training program, as described by Thibodeau:

> In 1963, under the Manpower Development and Training Act, the Northeastern Technical Committee of the A.P.A. [American Pulpwood Association] sponsored a Woodsmen Training Program. Cooperating with the above, besides members of the industry, were the Maine Department of Education and the Maine Employment Security Commission. Advance publicity of this first program was excellent, and on May 13th, thirty trainees were referred to the Maine Forestry School at Princeton by the Employment Service. Twenty-one finished the four-weeks course. Eighteen of these expressed a desire to work for members of the industry, ten to Great Northern and eight to others. On June 9th, eight reported to Great Northern at two of our camps in the Ashland area. One stayed two weeks, two stayed three weeks, and two others stayed five weeks and three stayed 15 weeks. Purchase and upkeep of used saws plus make-up pay for this group was in excess of $1,000.00.

So, what went wrong? Why didn't trainees stick with logging and become the region's future woods workers, the goal of the program? According to the author: "The first [of two] 1964 six-week session was held at the Northeastern Maine Vocational Institute, Presque Isle. This proved to be a poor location in that it was 35 miles from the woods training area, necessitating three hours daily travel. This program was geared for 35 to 40 trainees—only nine showed up and only one had the initiative, capabilities and willingness to go to work when five weeks' training was finished."[45] However, "poor location" was likely only part of the story, if it mattered at all. The second of Great Northern's 1964 sessions also went poorly. Although seventeen prospective students reported to the training site, only ten graduated, all of whom were offered logging jobs in the Ashland area. Only eight accepted, but all requested a two-week vacation before reporting for work. Unfortunately, when they finally did report, "when left alone they did not have the incentive, ambition, or desire to work," and Great Northern decided that "it was a waste of time and money to keep them any longer."[46] In the end, Thibodeau did little to elucidate strategies for the "effective recruiting of loggers," despite his article's promising title.

Wisely, George Carlisle from Prentiss and Carlisle, a forest management company located in Bangor, called a time out and decided to further investigate the dismal outcomes of this training. "After reading Mr. Thibodeau's paper, I wondered as to the thoughts and ideas of those who had attended these schools, what they were doing now, and as to how much wood they had cut," Carlisle wrote. "In an attempt to find answers, I obtained a complete list of the 58 people who had attended the schools in the past two years and sent each a very simple questionnaire . . . the following is a summary of the information obtained from the 38% who answered: (1) Only 50% were currently employed; (2) None were cutting wood, nor had they cut wood since completing the school; and (3) Those working were employed as follows: machinist; chicken farm (2); processing plant; paper mill (2); farm (2); saw mill; groom at race track; dish washer." When trainees were asked why they didn't pursue careers in logging, they responded with a litany of reasons, including "can't compete with cheap import labor; no future because the Canadians get the best chances; can't keep pace with the French; dull wearisome boring."[47] What Carlisle concluded was not that there was a shortage of available labor but rather that woods work was not an attractive employment alternative for many able-bodied workers and that no amount of training was going to change that. The industry was discovering the hard way that it couldn't train its way out of its logging workforce dilemma given underlying challenges of poor pay and remote and dangerous seasonal work. Perspectives from the western United States reinforced the need for more systemic changes if the logging industry were to bolster its workforce. According to one source in 1965, "The West, like the Northeast, has the problem in many areas of too short a work season and too small annual earnings to encourage loggers to stay," despite robust enrollments in logger training courses.[48]

A study by the Maine Department of Conservation, *Maine Woods Labor Supply, 1984–85*, appeared to concur, its authors suggesting that "the low status of the occupation, the low pay, and the backbreaking work combined to favor other employment," not a lack of logger training opportunities. They went on to write: "The drop in vo-tech institute applications points to a larger problem affecting woods harvesting throughout the state. This is the low esteem held for woods work. During the course of this research, comments were heard such as, 'Woods work is the lowest form of drudgery,' or 'If you don't straighten up, you will end up in the woods.' The poor image of woods work may account for the numbers problem that the vo-tech institutes are now facing. At a greater level, it is an important long-term problem for the forest industry with respect to attracting high-quality labor for woods work . . . The low regard held for woods workers is in contrast to the increasingly wide range of skills loggers must have."[49] What the authors failed to acknowledge was that the "low esteem held for woods work" was, at least in part, a function of low pay and conditions that caused loggers to quickly wear out—or to again quote Québécois logger Lucien Lachance from *Cut and Run*: "We work at a speed just to make a decent wage and at a certain age there will come a time, whether you are Canadian or American, when you won't be able to work this hard, to keep up this pace—it's impossible."[50] Fix the low pay and unsustainable pace of woods work and whatever poor image logging may suffer from may be mitigated, the anecdotes and the data would suggest. Nevertheless, the industry has insisted on using logger training and recruitment programs to fix logging's public image as a way of solving the shortage of loggers, ignoring what appeared to be the key determinants of the labor problem: low pay, poor benefits, seasonal work, and woods worker burnout.

Maine's northern forest was not the only place where logger training efforts were employed to solve woods worker shortages, instead of focusing on logging's underlying challenges. A 1970 study of logger recruitment in Michigan, for example, similarly concluded that "the labor shortage was not caused by an inadequate supply of rural manpower. During the period of wood shortages[,] reports were issued indicating a surplus of manpower levels in the very areas experiencing woods labor shortages . . . The generally acknowledged explanation is that alternative opportunities have been more attractive than woods work" and that more stable year-round work and employment benefits were needed to address the problem by elevating logging to a full-time profession, rather than simply a part-time enterprise punctuated by periods of unemployment. This could be achieved only by longer pulpwood contracts and better pulp prices, so that independent Michigan contractors could bank on more stable demand and prices that allowed them to cover costs and provide a margin for profit, given the considerable risk associated with investments in logging mechanization. Authors of the Michigan woods labor report concluded: "Regarding the possibility of training woods workers as a means of increasing the woods labor supply, it would be necessary to be able to offer such trainees a fairly attractive job. A program designed to train a person to be a pulp cutter is not likely to attract many potential workers under working conditions generally existing today . . . Emphasis must be directed towards improving general working conditions before considering training programs," which echoed what Carlisle and others realized a few years earlier.[51] In 1970, the income for a Maine logger was $5,504, whereas a construction worker made $7,057. The 1986 Maine woods labor study noted that "few full-time occupations in Maine posted a lower wage. Though the median income for loggers has doubled from 1970 to 1980, using the producer price index as a gauge, this reflects an actual income loss of 28 percent." During the same period, the number of loggers in Maine dropped from over 4,500 in 1970 to approximately 3,500 in 1983, some of which was likely due to increases in logging mechanization.[52]

Echoing refrains about the efficacy of logger training voiced elsewhere, a 1998 study of loggers from Georgia suggested that they preferred on-the-job training to more formal logger education that utilized professional trainers and that they were more concerned about retirement and health benefits than logger training. Although these employees expressed satisfaction with their work, especially logging's outdoor work environment, "neither logging contractors nor their employees currently rate training as very important in improving their productivity or job performance."[53] A year earlier, a study of nearly three hundred West Virginia loggers found that formal logger certification programs there were perceived by that state's logging community as too expensive, with an average certification training cost per worker of over $1,600.[54]

What did Maine's loggers think about how to attract and retain woods workers? A 2006 University of Maine study found that of 296 Maine loggers responding to the open-ended question "What could be done to attract people to the logging profession?, 145 (49 percent) wrote "improve pay" and 39 (13 percent) wrote "provide benefits." Only 13 logger-respondents (less than 5 percent) mentioned logger training, such as logging equipment training programs, vocational training, and apprenticeships.[55]

What can reasonably be concluded from over a half century of insights and initiatives related to logger recruitment? Despite forest industry efforts, the challenge of building sustainable logging capacity is less about a lack of training and more about the deeply rooted sense that logging is not a desirable profession, given its harsh working conditions, relatively low pay, a lack of employment benefits, and long periods of unemployment. Indeed, some who are unfamiliar with the knowledge and skills needed to be a successful logger or logging business owner have asserted that it's not a profession at all.

Reports of a shortage of woods workers continued to persist, despite shrinking markets and contractions in the industry. Much of the projected labor deficit appeared to stem from retirements of an aging workforce. In 2020 one headline read: "Logging Adds Over $600 Million to Maine's Economy, but the Industry Can't Find Workers."[56] At the same time, the Maine Department of Labor's 2018–28 employment outlook projected a 10.9 percent *decrease* in jobs for logging equipment operators, throwing the present and future condition of Maine's logging workforce into a state of some confusion.[57] The apparent disconnect notwithstanding, the persistence of logger training graduates in Maine's logging workforce continues to be a challenge. While a number of new recruits may give logging a try, very few appear to stick with it, realizing they can make more money and have employment benefits performing easier, safer, year-round work in some other, more stable sector of the economy.

In spite of conflicting projections about the state's logging capacity, the 2020 article referenced forest industry representatives who insisted that a key to the future of logging in Maine was more education, including initiatives like the Professional Logging Contractors (PLC) of Maine's Mechanized Logging Operations Program, a partnership with the Maine Community College System and forest industry that was launched in 2017. As the executive director of the PLC remarked, "We're slowly building a base," referring to the 29 graduates of the program in the past three years who all received job offers in the field before they even completed training. "While not all of them accepted jobs, the program has maintained a 75 percent retention rate in the industry."[58] However, it was unclear how many of the program's 29 graduates actually accepted jobs in Maine's logging woods. As with previous logger training efforts, proponents of mechanized logging equipment training were quick to point out the job opportunities, which are dwindling according to the Maine Department of Labor, but few appeared to be prepared to discuss the persistence of graduates in the logging workforce.

To further accentuate the point, in 2006 the *Bangor Daily News* reported that one of Maine's large forest management companies sponsored high school students enrolled in vocational forestry programs. As an incentive to hire company-sponsored students, the forest management company offered to reimburse logging contractors for some of the first six months of each graduate's pay. Of the seven students from across three Maine high schools who were interested enough to enroll in the program in 2005, four graduated from the program and were hired by logging contractors. Only one of the four was still working as a logger a few months later, a "25 percent success rate," considered "not bad" by a program spokesperson.[59] What happened to the other six students who originally enrolled in the program was not reported, nor was the persistence of the program's one success story beyond the first few months of work in Maine's logging woods.

Indeed, although education may be a key to training an able workforce in an increasingly mechanized and regulated industry, among the key findings of a 2019 PLC-commissioned study on the challenges facing the logging workforce were that wages of logging equipment operators ranked lowest out of 19 comparably skilled occupations and that since 2010 the average earnings in Maine's logging industry have increased by just over $2,100 per worker compared with an average increase of $5,500 for workers in industries such as construction, wood manufacturing, and pulp and paper manufacturing.[60] Logger training programs alone don't fix those stark realities, and loggers don't discourage their sons and daughters from working in the logging woods because there's a lack of training. Importantly, potential logging recruits understand that every mill closure in their state and in their town means a more challenging existence for those who, during better times, used to perform the difficult and dangerous work of supplying those mills with raw material. In addition, shifting the focus of logging capacity to logger training suggests that if only loggers were better trained, the industry's workforce issues would go away. The problem isn't pay or benefits; the problem is undertrained loggers, the industry would assert.

Who will log in Maine's woods? Loggers taking part in a group discussion in central Maine agreed that finding employees as well as the high cost of employee benefits forced them to work alone. None of the ten loggers participating in the discussion would recommend logging to their children. A group of ten loggers in northern Vermont, most of whom are also self-employed, one-man operations, agreed. They expressed concern about attracting workers to logging "at wages and benefits that workers expect elsewhere for unskilled labor."[61] Backing off the focus on woods workers from Maine and the United States and reinforcing the importance of a robust foundation of raw material suppliers to the health and sustainability of a forest economy, a report on the global forest industry suggested that any plans for sustainable forestry should include "adequate provisions for improvements in the social and labor aspects of forest management, specifically employment, income generation, working conditions, safety and training."[62] The authors questioned the sustainability of forest management in an environment without decent jobs and livelihoods. Whether a lack of "decent jobs and livelihoods" in Maine's logging woods helped explain the unraveling that was on the horizon in the state's forestry sector after decades of discord remains a matter of some speculation.

In addition to the challenges facing many logging businesses, Maine's logging woods were suffering from a poor self-image, an identity crisis, a lack of interest. Studies from the 1970s and 1980s showed that logging's occupational status roughly equated to that of railway porter and busboy, despite the skills and knowledge required to work productively and safely in the logging woods and manage a logging business. More recent studies have shown that along with working outdoors, loggers valued their independence, the very thing that some north woods Maine loggers appeared to be willing to negotiate away.

Those who proposed training as an answer to the question of who will log ignored the data, thinking that additional training alone would solve the problem of a shrinking workforce. Industry-sponsored efforts to boost Maine's logging workforce met with limited success, primarily because of the failure to address the underlying challenges facing many logging businesses, including low pay, poor job benefits, and lack of year-round work. Yet in the decades ahead, governors, state legislatures, and departments of labor would be undeterred, pouring hundreds of thousands of taxpayer dollars into logger training programs, ignoring the data and yielding instead to a still-influential forestry lobby, despite the growing irrelevance of the sector to the state's overall economy.

The thinking, too, was that with increases in logging mechanization there would be more young people interested in logging as a career. But while mechanization increased productivity per man-hour, logging continued to suffer from gaps in its workforce. After all, other industries were also becoming more mechanized, with equipment operators able to make more money and have job benefits closer to home in construction and other industries.

Ominously, studies of Maine's logging community, such as it was, also suggested that familial attachment to logging was breaking down. If the sons and daughters of loggers didn't want to log, in an industry that relied heavily on an intergenerational pipeline of workers, where was the next generation of loggers going to come from? A survey of Maine's general public suggested an aversion to following logging as a career path because of a perception of a harsh work environment and physically demanding work. And, at a time when many in the industry were whitewashing the increasingly dire condition of the state's forest economy, an informed Maine public was reading newspapers whose headlines proclaimed mill closures, border blockades, and work stoppages—headlines that suggested an industry in freefall, a "traditional industry" with a questionable future, despite expressions of optimism from the forestry sector's leadership, headlines of dysfunction and discord in Maine's logging woods, of an industry that could not seem to figure out whether it was at a crossroads or in crisis.

Chapter Seven
Crisis or Crossroads?

The fear that the paper companies are going to cut and run I think is a fear that's not based on fact. The companies have a tremendous capital investment here in the state and they're going to stay here to protect that investment. Mills cost hundreds of millions of dollars to establish; you don't run with those. They're in place and they're going to stay.

—Maine paper industry spokesperson Henry Magnuson, interviewed for the documentary *Cut and Run*, 1980

Paper mills employed about 5,200 Mainers at the end of 2015, according to the Maine Department of Labor, down from more than 15,000 in 1990. Just in the last three years, five paper mills across the state have closed.

—Christopher Burns, *Bangor Daily News*, 2016

Aside from the significant challenges associated with the uncertain future of Maine's logging workforce, other persistent dynamics conspired to threaten the state's forest products industry, including the conversion of private working forests to public parks, society's evolving forest values and forestland valuations, fleeting markets for forest raw material, aging mill infrastructure, unfriendly state policies and regulations, and global competition. Although comparisons are imperfect, experiences from other places in the region may provide some insight into the future of Maine's logging woods and the individuals, families, businesses, and communities that relied on it. Coming to grips with the turmoil created by the forest industry's culpability for its own unraveling, including a trail of misinformation and politicization, and the hardship left behind by mill closures is another matter. Maine's forestry sector was becoming "an industry of the past," dragging many of its working forests and forest-dependent communities along with it.

LESSONS FROM INSIDE A BLUE LINE?

Forestland ownership in Maine has been shifting, with fewer industrial forest owners and more nonindustrial private forest acres. As of 2016, 10 percent of the state's total forested acreage was publicly owned, with private and tribal ownerships accounting for the rest. Attempts to convert Maine's multiple-use working forests to single-use public preserves that would rely on an outdoor recreation economy have met with some resistance from the forestry sector, sportsmen's groups, and rural communities. For example, in 2011 a wealthy environmentalist and large Maine landowner advocated for the creation of a Maine Woods National Park in the state's Katahdin region, a rural, forested place with high unemployment and local residents who were, according to her, "tone deaf when it comes to the environment." Reportedly referring to Maine as a welfare state, she went on to describe the state as a having a "large population of elderly and obese people and an outdated forest products industry model." While arguably not a great approach if one was interested in developing partnerships with and endorsements from local communities—"She is trying to make a point by running the people down? It doesn't make sense," according to one critic—to her credit she also engaged with some key opinion leaders, including the head of a sportsmen's group, and was willing to donate over seventy thousand acres to the cause.[1]

Nevertheless, forest user groups, including representatives from the forest products and sportsmen's interests, along with the state's governor and legislature, helped scuttle the proposed national park plan, primarily over concerns about access to working forests and the resulting economic impacts on local communities. Pushback also came from the co-owner of Lincoln Paper and Tissue LLC, who questioned ceding so much land to the federal government to create a national park. Countering the characterization of Maine's "forest products industry model" as "outdated," he described the state's forest products industry in 2011 as efficient, facing a "tremendous increase" in the demand for wood that is driving concern about whether there would be enough loggers to harvest it.[2] Ironically, in 2015, just four years later and after a boiler explosion at his mill, Lincoln Paper and Tissue filed for bankruptcy protection, with hopes now pinned on reviving the mill site to manufacture cross-laminated timber as part of the Maine Forest Products Innovation Park. Additional mill closures across the state would soon follow, lending at least some credence to the

Maine national park advocate's criticisms of the region's reliance on a forestry economy. Coming up short of congressional support needed to create a national park, in 2016 President Obama signed off instead on the smaller, 87,563-acre Katahdin Woods and Waters National Monument, a designation that does not require congressional approval.

Lessons about the challenges related to converting private working forests to public forest preserves may derive from the New York State's Adirondack Park, a six-million-acre mosaic of private working forests and public forest preserve, punctuated by scattered rural communities and traditionally outlined on maps with a blue line. Reminding us of the creation of the Adirondack Park in the late nineteenth century, a failed attempt in the 1960s to turn the Adirondack Park into the Adirondack Mountains National Park whose boundaries excluded the estates of some wealthy Adirondack landowners, and the ultimate creation of the Adirondack Park Agency as the region's managing authority, the optics around Maine's national park approach may have been considered more paternalistic than collaborative, more promise than substance. Indeed, the Adirondack Park experience suggests that promises of employment and local economic development related to the creation of a forest preserve to replace otherwise multiple-use working forests might be viewed with at least some skepticism.

About the size of the state of Vermont and 87 percent forested, approximately half of the Adirondack Park is designated "forever-wild" forest preserve, a part of the park that is growing as the state acquires more and more private forestland to include in its preserve. Originally created as a rural land-use model that encouraged synergies between forest preservation and local community sustainability, the Adirondack Park has been losing population, the remaining, less mobile residents too often suffering from what one state official described as "poignant rural poverty." The poorly diversified Adirondack economy, hamstrung by land-use restrictions not found outside the park and tethered to seasonal, low-paying, and localized outdoor recreation businesses, has seen its forestry sector shrivel, challenged by relatively poor transportation networks, reductions in available raw material, and withering wood markets, some gone for good, others retreating to a better business environment outside the blue line, as I noted in my book *Adirondack Hard Times* (2021), about power and politics in the forests of the Adirondack Park.[3]

An example of the social and economic polarization that can occur when forestland passes from private to public ownership is the 2021 ruling that trees less than three inches in diameter may not be cut to expand snowmobile trails, or for any other reasons, on the Adirondack Park's forest preserve lands. The controversy pitted forest preservationists against New York's Department of Environmental Conservation, which had been engaging in the construction of Class II trails—those wide enough to safely accommodate snow machine traffic—on the preserve. Although some environmental groups hailed the ruling as a victory, struggling rural communities and sledders, who help sustain restaurants, motels, convenience stores, and gas stations through an otherwise economically bleak Adirondack winter, were hoping for a different outcome. The average snow machine rider spends an estimated $2,000 a year on snowmobile-related recreation, according to the International Snowmobile Manufacturers Association.[4] In 2011 the New York State Snowmobile Association partnered with the State University of New York at Potsdam to look at the impact of the snowmobile industry; the study estimated that it contributed $868 million to the state's economy. Of the snowmobile industry's economic impact in the Adirondacks, the report suggested that "it should be a reminder of the importance of public trails within the area and the role snowmobiling plays in the small communities of the Adirondack Park."[5] Snowmobilers are also important to local winter economies throughout the northern forest region, including the state of Maine. Snowmobiling was referred to as "the most essential winter staple in the [Aroostook] County economy" by a motel owner in the northern Maine town of Madawaska. "It's pretty much the only business we have," she asserted of an activity that reportedly brought hundreds of millions of dollars to the state's economy every year and supported thousands of rural jobs.[6] Accusations of elitism and classism echoed through the Adirondacks, falling on deaf ears among an environmental glitterati that appeared insensitive to the region's already struggling working class.

So far, except for some theme-parked places such as Lake George and Lake Placid, the promise of a vibrant recreation economy in the Adirondack Park has been illusory. According to an editorial in the *Lake Placid News* in 2020 that presented arguments in opposition to the state purchasing a private 36,000-acre estate that would have been added to the park's forest preserve, local Adirondack communities have realized few, if any, economic benefits from such designations in the past, citing the 18,400-acre St. Regis Canoe Area, designated in 1972 as part of the Adirondack Park State Land Master Plan, as an example.[7] The more recent 2016 inclusion in the park's preserve of the Boreas Ponds Tract has, so far, not brought an economic boost to the surrounding Newcomb area, the editor argued. Other reasons mentioned for not purchasing the estate and designating it forest preserve included the already strapped Department of Environmental Conservation, the state agency charged with regulatory authority on environmental issues, and overuse of the park's preserve where "there are just too many people on the trails," the state's purchase and designation of its lands as preserve "doing more damage than protection."[8]

Not only has the contention of outdoor recreation as economic savior in the Adirondack Park not panned out, but the near exclusion of other sectors of the economy has disadvantaged local communities. A study published by the North East State Foresters Association reported that, statewide, forest-based manufacturing contributed $6.9 billion in value of shipments in 2005, whereas forest-based recreation generated only $1.9 billion. In addition, the same report claimed that forestry and logging provided 57,202 jobs and a payroll of over $2.1 billion in the state as a whole, while forest-related recreation brought 14,600 jobs and a payroll of $300 million, reflecting average salaries of $36,712 and $20,548 for forest products and outdoor recreation employment, respectively. Importantly, the study stated that one thousand acres of forestland in New York supported 3.0 forest-based manufacturing, forestry, and logging jobs and only 0.8 forest-related tourism and recreation jobs.[9] A 2014 report noted that "many tourism positions are poorly paid, part time and/or seasonal, offering few benefits . . . average weekly wages of $255 . . . with many of the jobs going to low-skilled workers and youth, especially in the summer," the Adirondack region's peak tourist season.[10] And despite the broader, Adirondack Park–wide economic benefits of tourism, remote communities, once reliant on the forestry sector, have generally been left behind.

The degree to which the Adirondack Park experience would predict similar outcomes in Maine's north woods is, of course, unknown. But writing about the Maine Woods National Park proposal in her 2020 book

Mill Town, author and Maine native Kerri Arsenault asserted that some "opponents were concerned a national park would hurt industries and families that produced Maine's forest-related products: lumber, pulp, furniture, paper, and wood, which comprise the state's largest exported goods. And with five paper mills closing between 2008 and 2016, leaving approximately 2,500 folks unemployed, a national park could ruin livelihoods of people halfway ruined themselves." Not unlike the conflicted attitude toward people "from away" held by many local Adirondackers in their role as both stay-away complainers and welcome-wagon enablers, Arsenault wrote that native Mainers "lived with a tolerant disdain for tourists, feeling both used by and reliant on them, a conundrum we've yet to solve; it was like our codependent relationship with the forest itself. Tourists came, enriched us, then left us with a tenuous economy at the mercy of Vacationland whims. Also, their infrequent visits to their (often) second homes made town centers as sleepy as the dead ache of frostbite," similar to the experience of many tourist-dependent Adirondack towns during the nonsummer months.[11]

Voices from elsewhere in the northern forest region have also cautioned about bureaucratized rural land-use models like that found in the Adirondacks. "Centralized planning may be considered by many as successful, but for the rural landowner and worker it is usually devastating," wrote one Vermont opinion leader, referring to the idea of land-use regulation in the Northern Forest region of Maine, New Hampshire, New York, and Vermont, similar to that found within the Adirondacks' blue line. "We need only look at the Adirondack Park to see what can occur," he continued. "Nowhere else in the Northern Forest has so much hostility developed between governmental entities who administer land-use controls and the residents of the area. People see their property rights more and more limited, and they resent the continued use of public funds to buy land that will be forever removed from the production of commodities."[12] That many national monuments have been preludes to the ultimate creation of an expanded national park, including Maine's Acadia National Park, which had its origins as Sieur de Monts National Monument in 1916, may serve as a note of caution to those who are reluctant to transition private forest ownerships to state or federal jurisdictions.

HEADING SOUTH?

There may be parallels between the evolution of Maine's forestry practices and logging workforce and the current logging woods realities in nearby southern New England. Although more heavily populated, the states of southern New England—Connecticut, Massachusetts, and Rhode Island—are primarily forested, providing forest-related services and timber and nontimber forest products as diverse as outdoor recreation, wildlife habitat, open space, sawlogs, and firewood. Massachusetts, the eighth most forested state in the United States, is 62 percent forested, and 60 percent of Connecticut and Rhode Island is forested. Similar to the states comprising northern New England, most of the forestland in the region is owned by nonindustrial private forest owners, but three-quarters or more of southern New England forest ownerships are less than ten acres in size.[13] Landowner objectives in the region are diverse and less inclined to favor timber harvesting. A 1992 University of Massachusetts study found that woodlot owners in southern New England more often owned forestland for nonconsumptive uses such as wildlife observation and aesthetic enjoyment than for more consumptive uses, such as income from timber removal.[14] As a result, despite a landscape dominated by forests, southern New England is a net importer of forest products. Massachusetts, for example, produced 3 percent of the wood products that it consumed, and its forest products output ranked twenty-seventh in the country, according to a 2010 study. In addition, forest cover in the region has been decreasing because of population density and development pressures in the region. In Massachusetts, reductions in both logging capacity and logging opportunities on state land appear to further challenge the wood products industry there.[15] Regulations driven by shifting public opinion about logging and rural land use have also impacted forest practices in the state. "Today, about 40,00 acres, or 13 percent of the forests and parks managed by the state Department of Conservation, are off-limits to logging," the *Northern Logger* reported in 2010. "Under the new plan, logging will be banned on at least 185,000 acres, or 60 percent of the lands. In most cases, clear-cutting will be limited to one-third of an acre," reflecting "the growing use of public land in Western Massachusetts for recreation, and was adopted after an outcry about logging practices on some of those properties."[16]

Unlike Maine's north woods, southern New England is characterized by large population centers and densely populated nonmetro areas. A half century ago, one observer described "megalopolitan" southern New England as "composed of forests interlaced with belts and pockets of cities and suburbs that sprawl into agricultural lands."[17] Indeed, according to a 1992 study, among the key challenges to the sustainability of forests and forest products economies in Massachusetts were changes in land use and the attitudes and values of the region's forest landowners.[18] Woodlot sizes have also been shrinking as forests become subdivided because land values far exceed their potential to generate income from forest management, with woodlot sizes in Massachusetts averaging less than eleven acres in 1996—a quarter century ago.[19] When asked about the challenges faced by loggers in Massachusetts, a 2011 study quoted one logger participating in a focused discussion as asserting that "the [small] size of woodlots is a problem," especially given the high cost associated with moving mechanized logging equipment. He added that poor weather that does not permit harvesting combined with the state's logging regulations have "turned [logging] into a part-time gig."[20]

The situation does not appear to be improving. According to one estimate, over forty acres of forests and fields in Massachusetts are lost to development every day. Asked about timber sale acreages below which they would not bid, loggers indicated that they would harvest woodlots as small as five acres and 17,000 board-feet but would not drive more than twenty-five miles to do so, and that at least 30 percent of the timber on woodlots of this size would have to be of high quality for them to be interested.[21] Supporting the challenges to timber harvesting associated with population pressures in the region, a 2010 University of Maine study found that 86 percent of wood procurement managers from Connecticut and Massachusetts who responded to a survey saw too much development and shrinking woodlot sizes as challenges confronting their sawmill businesses, while 69 percent indicated that too much sprawl was a barrier.[22]

In addition to the fragmentation of forests by roads, power lines, and development, among the most important challenges to a sustainable forest products economy in Connecticut were parcelization, negative public opinion of logging, and concerns about a diminishing logging workforce, challenges that are likely pervasive throughout the southern New England region. State and local regulations have also challenged loggers from Connecticut. During a meeting of seven loggers from Connecticut in 2005, one logging business owner observed that "each town has its own regulations . . . It took me two years to get a wetlands permit." On the public's view of loggers, another participant claimed that "there are only about 500 loggers left in the state . . . If we disappeared, no one would care."[23]

To better understand these challenges from the perspective of the region's logging community, a 2011 University of Maine study analyzed survey responses from 252 professional loggers in the southern New England states. Loggers from the region cited day-to-day operating costs, equipment and insurance costs, the price of stumpage, a shrinking forestland base, and harvesting regulations as challenges to maintaining or expanding their businesses. The study also found that loggers in the region had a diminished sense of occupational prestige compared with loggers from New England's northern forest. Survey results and post-survey focus group comments suggested a southern New England logging industry confronted with several challenges, including a general perception of the public's lack of respect for loggers and logging and a disconnect between forest products and the work that loggers perform. One group participant sensed that, because of the overregulation of logging, "they don't want timber harvesting in the state of Massachusetts"; a Connecticut logger stated that "a lot more towns don't allow [logging] work on Sunday . . . I have to work when the weather is good for logging . . . There has to be a balance between regulation and reality on the ground." The meeting was concluded with this pessimistic comment about the future of logging in the region: "Slowly but surely they [the government and the public] are taking the joy of the job out of me . . . I wonder if we are at the end of the road and we're no longer a big enough voice."[24]

There is some indication that loggers from southern Maine share a sense of occupational choice and prestige that is closer to that of southern New England loggers than to that of their northern Maine counterparts. Summary results from two studies showed closer agreement between southern Maine loggers and southern New England loggers on responses to the statement "I became a logger because" than between southern Maine and northern Maine loggers. For example, 9 percent of southern Maine loggers and 8 percent of southern New

England loggers, respectively, disagreed with the statement "I could not find another job nearby," while 21 percent of Maine's northern county loggers agreed. Similar results were reported for the statement "It was the best paying job available": 16 percent of southern Maine loggers and 12 percent of southern New England loggers agreed, versus 28 percent of northern Maine loggers. Fewer southern Maine loggers (16 percent) and southern New England loggers (12 percent) indicated that they logged because they "don't have the education for a different job" than northern Maine loggers (28 percent). Considerably more northern Maine loggers (65 percent) than loggers from other regions (southern Maine, 45 percent; southern New England, 35 percent) said that they became loggers because they "come from a logging family," accentuating some similarities in education and familial attachment to logging between southern Maine loggers and southern New England loggers compared with those from northern Maine. While one might anticipate greater employment opportunities in southern Maine and the southern New England states than in more rural northern Maine (where a cascade of pulp mill closures and contractions was well underway), the lack of employment agility of workers there due to comparatively lower education rates, combined with a stronger sense of intergenerational attachment to logging, suggests a northern Maine whose forestry sector—and rural economy generally—was spiraling out of control. In addition, while over 60 percent of loggers from all three regions agreed that "the public views loggers as unskilled," reactions to the statement that "(t)he public respects loggers and the work they do" showed considerable divergence across the study regions.[25]

What does the southern New England experience predict for Maine's logging woods? A 2007 article reviewed the impacts of population pressures on Maine's loggers and logging businesses. Almost 700 loggers and 71 procurement managers from the state responded to a survey. The authors found that almost two-thirds of Maine loggers surveyed agreed or strongly agreed that more woodlots would be harvested for development and approximately 85 percent agreed or strongly agreed that logging chances would be smaller. Almost one-half agreed or strongly agreed that there will be less logging in their area because of sprawl. The study asked similar questions of wood procurement managers in the state. When questioned about the percentage of acreage or volume from which they purchased wood in 2001 that was from so-called terminal harvests (forestland being harvested for the last time because of subsequent planned development on those lands), the average was 11 percent. When asked about possible barriers to maintaining or expanding their businesses, 42 percent said that sprawl was an important or very important problem, 30 percent said that too much development was a barrier, and 28 percent cited woodlots that were becoming too small to harvest economically. However, much higher percentages of respondents indicated that stumpage prices that were too high and harvesting-related regulations had a more negative impact on their wood supply than factors related to population and land use. If results of the study estimate future conditions, the authors concluded that many of the state's loggers will have to travel further and move their equipment more often for logging opportunities, increasing their production costs. If these costs are absorbed by loggers rather than by wood consumers, additional erosion of logging capacity would likely result, the authors concluded.[26]

And land values in Maine are shifting, especially in the state's southern tier, where the city of Portland was among the top-ten fastest urbanizing metropolitan areas in the United States, according to a 2001 report.[27] An astute observer of the Maine forestry experience asserted that "the land market is saying that it is not financially feasible to grow wood in much of the Maine forest, as the land is too expensive."[28] So far, it appears that forestland conversions in southern Maine have been offset by the reversion of farmland to forestland, and the amount of forestland statewide apparently remained relatively stable from 1958 to 2008.[29] According to one source in 2010, however, for the first time since agricultural abandonment began in the mid-1800s, New England (excluding Maine) is losing forest area."[30] At the same time, the terminal harvesting of private forestlands has been increasing in Maine, potentially signaling a collision of land uses, especially in the state's southern tier, with agricultural and forest lands the ultimate losers.[31] If this trend continues, like their southern New England counterparts some Maine loggers will be forced to travel further and compete more keenly for stumpage, driving up their wood procurement and transportation costs.[32] Others may transition from full-time logging businesses to land management companies that provide services to nonindustrial private owners who prioritize noncommodity forest values (e.g., viewsheds) over growing timber, not unlike some of their neighbors in southern New England.

COMING CLEAN ON BIOMASS ENERGY

Woody biomass has been an important market for low-grade trees and so-called waste wood from forest harvesting and processing in the northern forest. As with the manufacture of other forest products, loggers have been on the front lines of raw material extraction to meet biomass demand, often investing in expensive mechanized equipment—feller-bunchers, grapple skidders, whole-tree chippers—to more efficiently produce whole-tree chips and low-quality logs for the biomass market.[33] But after an initial surge in Maine's biomass industry in the 1980s and 1990s, the biomass sector began to unravel, its future as a stable market and an alternative, renewable energy source threatened by fluctuations in prices for fossil fuels and other energy sources, global biomass markets, and science that questions the forest, environmental, and human health impacts of biomass removal, transportation, and cogeneration.

One observer of the region's biomass scene in 2016 stated: "Biomass electricity has a tough road ahead of it—and the loggers who supply these facilities need to think hard before the next (whole-tree) chipper purchase. Absent some real policy intervention or markets rebounding in unpredictable ways, the biomass market in the northeast will be much smaller in three to five years."[34] The following year, Maine attempted to take some of the risk out of significant investments in the mechanized logging equipment needed to harvest, skid, and chip low-quality whole trees efficiently. Under the 2017 op-ed headline "As Loggers Get Stiffed, Maine Learns a Lesson about Propping Up Struggling Industries," the contributor wrote about Maine's $13.4 million "bailout" of the state's biomass industry. The aim was "to keep Maine loggers working to provide wood to biomass plants, ultimately forcing residents and businesses to buy electricity produced at the biomass plants at above-market rates."[35] Essentially, biomass producers would receive subsidies in exchange for keeping wood suppliers in business for two years. For some critics of the state's 2016 biomass bailout legislation, the agreement reduced to diverting millions of taxpayer dollars to biomass producers. However, the subsidy also signaled the state's support of both its logging businesses, rural communities, and homegrown, renewable fuel. Reportedly, the controversial arrangement did not work exactly as planned, at least for some of the loggers, to whom the publicly subsidized biomass companies allegedly missed or made late payments.

A comprehensive 2017 study of Maine's biomass industry, "Maine Bioenergy at the Crossroads" summarized the biomass subsidy and environmental situation this way: "Taxpayers and ratepayers have doled out over a quarter of a billion dollars since 2008 in state and federal subsidies and grants, aimed at keeping Maine's biomass industry afloat, in a desperate bid to save forest industry jobs across the state. The payments have done little to stop the bleeding from an industry that generates electricity too dirty to be eligible for clean subsidies in some neighboring states, and too expensive to compete with alternatives without multimillion-dollar subsidies." The study was also critical of what it called the "industry studying itself" phenomenon, where Maine's matters of forest economics have been too often addressed by a commission or study group comprising the usual suspects from the forest products industry who have a vested interest in any outcomes or recommendations.[36]

The previous year, a *Bangor Daily News* editorial assessed biomass subsidies this way: "On its face, it seems like a simple solution: If Maine electricity users pay a bit more each month, Maine loggers can keep their jobs. It's not that simple, of course, which is why lawmakers are struggling with a bill that would prop up the state's biomass plants in the hopes of protecting logging jobs." Acknowledging that biomass energy plants "remain important customers for Maine's many timber companies and contractors," the editorial also asserted that "there are no guarantees the biomass plants will remain open," citing several examples of Maine biomass plants that were already in various stages of being "shuttered." Indeed, two days before it published its editorial, Mainebiz ran a story titled "Loggers Suffer One-two Punch with Mill and Biomass Plant Closures," in the wake of the closure of pulp and paper mills in Madison and Jay, as well as biomass power plants in West Enfield and Jonesboro, with further biomass facility closures threatened. The article quoted a highly regarded Maine logger as lamenting, "You try to plan for market conditions but you have no idea what the weather conditions will be when the markets are peaking, or if the market demand will be strong when the weather conditions are ideal."[37] The *Bangor Daily News* editorial was also critical of a proposal before the state's legislature that referred to "biomass as an energy source that produces 'zero greenhouse gas emissions.'"[38]

Indeed, there have been other suggestions that any initiative to prop up struggling biomass markets that portrays woody biomass as environmentally friendly should be viewed with at least some skepticism. Biomass has been heralded by some as a green alternative to fossil fuels, yet the reality is much more complicated and the science much less settled. It can take a long time—decades—for a replacement forest in Maine to reach the carbon absorption levels of a forest that was harvested and subsequently converted to woody biomass–fueled energy. Soil scientists are still exploring the degree to which forest biomass removal impacts the soil's ability to store carbon, as well as the impacts of whole-tree harvesting, the typical logging method for producing biomass, on short- and long-term soil quality. Moreover, some scientists argue that the combustion of woody biomass not only produces pollution that could create or exacerbate an array of health problems, including asthma, but also releases carbon dioxide at levels greater than those for the burning of coal, since wood burns less efficiently because of its higher moisture content. Importantly, woody biomass as a renewable energy source doesn't eliminate the need to burn fossil fuels to mechanically harvest and transport a relatively low-value forest commodity like wood chips or low-grade logs. And as the region's pulp, paper, and biomass facilities continue to shutter, trucking distances to the next nearest low-grade markets lengthen, requiring a greater reliance on fossil fuels to transport forest biomass raw material.

While woody biomass is renewable when managed sustainably, biomass derived from land clearings for conversion to some nonforest use can negate this, especially as forestland owners and land developers witness the relative ease with which mechanized biomass logging operations can clear land, create scenic views, and improve bare land values, while returning stumpage revenue for the tonnage removed. And an institution's or state's well-intentioned commitment to purchasing woody biomass only from forests that are certified as sustainably managed is not likely to deter some unscrupulous suppliers from chipping trees that should be set aside for higher-value forest products, such as sawlogs, thereby potentially reducing their value and the stumpage fees paid to landowners. In other words, forest biomass may not always be derived from so-called forest waste wood—otherwise nonmerchantable parts of trees, such as branches and tops—as many biomass proponents would like the public to believe. Monitoring the sources of woody biomass raw material, from the stump to point of delivery, will require a commitment from the industry and state and local governments. According to a participant in a 2005 group meeting of five loggers from nearby Vermont, "I've never been asked by any mill for proof of certification" when delivering raw material to a sawmill that ostensibly bought only logs that carried some form of supply-chain certification. It may be a mistake to assume that this doesn't occur in what's left of Maine's biomass sector.[39]

None of this implies that woody biomass should be written off as an alternative, renewable energy source in Maine or elsewhere. However, it does suggest a responsibility on the part of the forestry community and woody biomass supporters to "come clean" on the economic, environmental, and forest and human health impacts of woody biomass as an energy source and their strategies for mitigating those impacts. It also calls for forest scientists to decouple themselves from unhealthy alliances with the forest industry that may influence their research methods and a dispassionate interpretation of their research findings.

What is the future of woody biomass in Maine and throughout the United States? "Every time it looks like wood fuel has a bright, warm future in this country, the price of oil takes a dive, throwing the biomass business into confusion and chaos. This pattern has been repeating itself on a regular basis for at least four years, and the confusion and chaos are back . . . The emerging biomass fuel market was counting on continuing high prices to keep wood fuel competitive with oil and gas," one observer asserted about the industry in 2015. Describing the often risky investments made in a sector that relied in large part on relatively high and more stable fossil fuel prices, he continued: "The United States has . . . an established wood energy infrastructure—much of it built in recent years during periods of high fossil fuel costs and the prospects of future shortages. It consists of biomass-fueled power plants, pellet and paper mills and a large production capacity in the woods in the form of whole-tree chippers and whole-tree production systems"[40]—"systems" that have transitioned many of Maine's logging contractors from petty producers to owners of highly capitalized businesses. For some of the state's loggers, the biomass crash resulted in idle capacity and the debt that follows close behind.

Despite dramatic losses in markets for low-grade trees, including for pulp and biomass, Maine's most recent forest industry initiative, referred to as FOR/Maine (Forest Opportunity Roadmap/Maine), aspires to grow the forestry sector's contribution to the state's economy by 40 percent, or $12 billion, by 2025, according to its 2018 report.[41] Given the steep downward trajectory of the industry both leading up to and since the publication of that report, especially in the pulp, paper, and biomass sectors, the goal seems extremely ambitious.

"LEARNING TO LIVE WITH THE SILENCE"

"Your world is torn apart. Part of you is gone." That's what Millinocket town counselor and former mill worker Mike Madore said about the closure of his town's paper mill.[42] "When the noise [of the mill] went away, the jobs went away, the money went away," he lamented. During the industry's heyday from the 1950s through the 1970s, "when demand was high, eleven pulp and paper mills operated throughout the state," employing some 18,000 Mainers, according to a 2017 WGME radio report. "Things changed in the 1980s," with only 4,000 jobs in the industry by 2016 and six mills still left operating in the state. Yet, despite the steep downturn, the state's forest industry remained "a big economic machine," claimed the Maine Forest Products Council. "We're still very much alive and kicking and a growing kind of industry."[43]

A growing kind of industry. According to one estimate, between 1990 and 2020 "Maine lost 13,000 out of 17,700 jobs in the papermaking industry (about 73 percent), with about 2,000 of those lost since 2013."[44] In a 2016 article titled "Why Is Logging Dying," another source estimated that "in 1980, there were 25 pulp and paper mills in the state. Today, two-thirds of those mills are gone."[45] What happened to one of Maine's most iconic industries? The industry and Maine's governor blamed mill closures on high energy costs, global competition, and the paperless digital revolution. Although those external factors surely contributed to the industry's misery, it was more complicated than that. "In the early 1990s, concerns about the competitive position of Maine's papermakers were already in the air. The mills were aging and often in locations where expansion was difficult. The timber supply was tightening; mills were automating individual processes or closing uneconomic machines," wrote Maine forest economist Lloyd Irland. "Large modern mills appeared in the South in the 1950s and offshore after the 1990s. The largest world-class mills were as much as ten times larger than the typical Maine mill. In capital-intensive and energy-intensive industries like paper, plant-level scale is critical. Nationally, output per worker, according to US Bureau of Labor Statistics Data, nearly doubled between 1990 and 2015, significantly reducing mill headcounts per ton produced."[46] Maine's plodding forestry sector just couldn't keep up.

Nevertheless, despite credible, data-informed assessments and undeniably grim realities, the state's forest products industry appeared to remain in denial. The 2005 Governor's Council on the Sustainability of the Forest Products Industry, acknowledging the challenges presented by global competition, asserted that "Maine's forest products industry has long been an anchor of Maine's economy, providing well-paid jobs, community stability and support for traditional land uses."[47] While considering the state's forest products industry as "an anchor of Maine's economy" has been a bit of a stretch for several decades leading up to the council's 2005 report, and many loggers would argue whether logging jobs were ever "well-paid," some of the council's assertions were about to seriously unravel.

At the start of the recession that began in December 2007, Great Northern Paper Company closed its Millinocket mill, once hailed as the largest paper mill of its kind, with the town projected to lose almost half of its population by the 2020 census as a result. A year later, the regional manager of the Forest Resources Association (FRA) declared that the state's forest products sector was being challenged by "some of the worst economic conditions in 40 years."[48] Yet despite recent paper mill closures in the state and the dire situation described by the FRA, in 2013 the Maine Forest Products Council, while acknowledging economic challenges ahead, proclaimed that the state's forest products industry "has a bright future" and that "Maine's forest economy has developed so quietly that many of our neighbors don't realize how much our industry has changed."[49]

Continued grim news was just around the corner for Maine's forestry economy, however. In 2014, the year following publication of the curiously upbeat Maine's Forest Economy report, Verso Paper announced the imminent closure of its Bucksport Mill, impacting the jobs of some 570 mill workers, and the Great Northern Paper mill in East Millinocket filed for bankruptcy, following a layoff of 212 workers.[50] In 2015, the Old

Town pulp mill announced the suspension of its operations, affecting 195 workers, and Lincoln Paper and Tissue filed for bankruptcy after a boiler explosion two years earlier from which it couldn't fully recover, impacting 128 employees.[51] The bleak news for the state's forest products industry continued in 2016, as Madison Paper Industries closed, eliminating another 214 jobs.[52] As of 2017, only five pulp and paper mills—Sappi Fine Paper in Westbrook and Skowhegan, Pixelle Specialty Solutions in Jay, Catalyst Paper in Rumford, and Woodland Pulp in Baileyville—were still functioning in the state.

"Turmoil has marked the past year in Maine's pulp and paper industry," according to one astute observer in 2016, "and it's not over yet. While there has been investment, stabilization and innovation at some mills, the headline is certainly the loss of markets—many permanently." Of the closure of the Verso mill, he asserted that "this mill was later sold for scrap and demolished. Paper won't be made there again." Outlining some of the broader ramifications of the closures, he continued: "For those on the wood supply side, including landowners, foresters, loggers and truckers, 2015 was a rough year . . . Maine lost markets for more than two million tons of pulpwood and chips. Some of these facilities also had associated biomass markets. These are gone as well . . . Already we have seen some loggers returning equipment to the dealer, in essence walking away," speculating that "by mud season 2016, the state would see a 10–15 percent reduction in logging capacity."[53]

A few months later, in 2017, and further signaling the industry's ongoing struggles, the Maine Pulp and Paper Association, a trade organization with twenty-one supporting members, folded. In existence for half a century, the association cited the recent closure of six paper mills and two biomass power generation plants to explain a $1.3 billion decline in the industry's economic impact between 2014 and 2016 and the loss of some five thousand jobs, despite upgrades and expansions in a couple of the state's remaining paper mills.[54] The closed mills cited high energy costs and decreased demand for their products, representing "an economic crisis of unprecedented magnitude," according to Maine's two U.S. senators.[55] Importantly for landowners and loggers, the closings slashed markets for low-grade trees and logs. A vanishing market for low-grade wood also challenged the forestry sector's ability to conduct improvement silviculture and certain types of regeneration harvests in the region's forest stands without relying on out-of-pocket or state and federal subsidies. Or as stated by an observer of the region's pulp and paper industry and put in way that frames the challenge of mill closures to more closely reflect the reality of logging businesses, with the closure of pulp mills in Bucksport, East Millinocket, Lincoln, Old Town, and Madison, "the (pulpwood) market has shrunk by 275 (truck) loads (of pulpwood logs) per day, every day . . . For lost markets in Maine, most of those mills are now scrapped and won't return under any circumstances."[56] Since that bleak assessment, four of the state's six biomass plants have closed, further challenging mill-dependent towns, families, and businesses and the state's ability to improve its working forests by removing low-grade trees. Unfortunately, Maine's forestry sector continued its freefall in 2020, with Sappi's Westbrook mill announcing the layoff of seventy-five workers, 30 percent of its workforce, and Pixelle laying off fifty-nine workers as the result of an explosion at its Androscoggin Mill.[57]

What is the future of all the infrastructure left behind after these closures? According to a 2019 *Bangor Daily News* report, repurposing of mill infrastructure and redevelopment of areas impacted by the closures have been uneven, with many promises of investment either slow to develop or not materializing.[58] For example, the East Millinocket site was ultimately sold to a company from Florida specializing in "demolishing obsolete properties"; much of the Bucksport mill has already been demolished, with plans for developing a land-based salmon farm, among other initiatives.

What does all of this mean for Maine's loggers? According to one estimate, between 2014 and 2016 alone, the state saw a 20 percent decline in the tonnage of wood harvested, mostly from small and low-grade trees. During that period, the combined loss of pulp and biomass markets has been estimated to result in a contraction of the state's logging workforce of at least 15 percent.[59] "There's a lot of wishful thinking going around, but I don't think the woods industry will ever be the same," lamented one area logger in the wake of the explosion at the Androscoggin Mill in Jay in 2020.[60] In addition, despite optimistic assessments and projections about the "solid wood" sector of the state's forest industry (higher-value sawlogs for construction, furniture, and flooring, for example) on the part of some forest industry advocates, closure of pulp and paper markets will also negatively impact sawmill businesses, because it reduces the ability to sell wood chips generated by milling processes, a significant source of sawmill revenue. In 2015, one Maine sawmill owner noted that "what I see is a real crisis on our steps and it's going to get worse if we don't do anything . . . The real loser [amid the closure of Maine's pulp and paper mills] is the small guy with a skidder and a chainsaw . . . If we lose the pulpwood and sawmill industry, you just can't start it up again."[61]

How did closures impact local mill-dependent communities, places that existed primarily to support a single wood-consuming business in a remote, upcountry place devoid of much other opportunity? Using the closure of the Great Northern mill in Millinocket as an example, one observer wrote: "The once mighty Great Northern Paper has closed for good. The mills have been sold through bankruptcy proceedings and are now being scrapped. A significant number of loggers, landowners, truckers and others were owed significant sums by these facilities, and they have little chance of recovering their losses . . . The shuttering of Great Northern mills—and the finality of it—has been a devastating economic and psychological blow to Millinocket and the surrounding communities. These are towns, after all, that were built to serve the mills. While their future is far from certain, what is clear is that paper manufacturing won't be part of the region's future."[62] According to Irland, "In just a quarter century, an industry once known by its superlatives—biggest mills, fastest paper machines, biggest dams, biggest land ownerships, and biggest logging machines—saw most of its manufacturing capacity, employment, and payrolls pass into history." He added, "It left small communities at the edge of the Maine woods with vacant industrial sites, dashed hopes, and painful adjustments ahead. Maine's mills saw demand for their products collapse rapidly," in an industry and in a region too often characterized by a lack of economic diversity, resilience, and agility.[63] And it's not simply a loss of jobs. The closures of paper mills can also drain a place of community, leaving social dislocations that often follow high rates of poverty and unemployment. Describing what's sometimes left behind when mills pull stakes, Kerri Arsenault wrote, "The opioid crisis is bludgeoning my hometown. Our special education population is the highest in the state. Rumford also has the highest rate of crime. And on top of it all, our environment is like a garbage can," a reference to the polluted, rivers, groundwater, and soil left behind by some papermaking processes—a legacy of not just polluted natural resources but also contaminated and shattered communities.[64]

Writing about the closure of the Verso Paper mill in 2014, the *Northern Logger* stated that "when plants such as the one in Bucksport close, entire communities, where families may have been employed for generations, are often devastated."[65] According to town officials in Bucksport, "property taxes from the mill made up about 44 percent of the town's approximate $12 million annual budget."[66] An article in the *Boston Globe* in 2020 reported that before its explosion, Jay's Androscoggin Mill had delivered at least half the town's tax revenues,[67] demonstrating the risks associated with a community or a region tethering its future to poorly diversified economies. While that may have worked decades ago, with generation after generation of Maine workers confidently relying on the mill for employment and community stability, global competition, fluctuating fuel prices, aging mill infrastructure, forestry sector dysfunction, and shifting forest values have combined to create new and less economically durable realities.

Also challenging are the impacts of the loss of pulp, paper, and biomass markets on forest health and stewardship. The removal of low-quality timber that typically represents the raw material for these markets predicts fewer opportunities to practice improvement, sanitation, and salvage silviculture in Maine's forests. One Maine logger noted in 2014 that "sadly we're having to work around a lot of timber that we should be harvesting if we had the [low-grade] markets to do it. When it comes to biomass, everyone's really operating at about half capacity or maybe even less, and there are a lot of chippers parked, and the ones that aren't parked are pretty much chipping part-time. As a result, what we're doing in the woods right now is not always the kind of forestry I would like to be doing; we're going around [low quality] trees that we should be cutting and having to do more high grading just because of what we can sell," suggesting a devaluation of some of the state's timberlands.[68] And that was before the loss of several additional pulp and biomass facilities in the state. Because the occasional sale of stumpage by industrial and nonindustrial forest owners helps pay the taxes on their lands while providing some revenue, dwindling markets for what they grow predict turnover and instability in forestland ownership in the state, with no guarantee that long-term forest stewardship will be an objective for the next owner, further encouraging terminal harvesting and forestland parcelization to compensate for the costs associated with forestland ownership. In addition, once logging businesses transition to more highly mechanized and more capitalized equipment to accommodate the removal of vast amounts of low-grade trees, when those low-grade markets deteriorate there can be a tendency for the logging equipment extant in a region to drive silvicultural decisions.

Certainly, there have been some signs of life scattered across Maine's forest industry landscape over the past several years. On the brighter side, the Old Town mill has seen a new owner resume pulp, but not paper, production, and there have been equipment upgrades in a couple of Maine's surviving mills. In 2016, for example, Sappi North America announced a $25 million investment in its Skowhegan mill. Recent investments in other mills in the state, including $120 million in the St. Croix Tissue mill in Baileyville and $12 million in Twin Rivers Paper Company mill in Madawaska, have resulted in improved employment potential in an otherwise moribund industry. But despite significant investments in the modernization and production capacity of some of Maine's remaining paper mills and among the independent loggers who supplied them with raw material, employment in the state's forestry sector as a whole has continued to decline, while misinformation and misleading optimism about the state's industry have persisted.

UNE CRISE FORESTIÈRE

Québec's forestry sector has faced similar challenges in the twenty-first century, with reverberations inevitably felt south of the border in Maine's north woods. In 2004, for example, Québec's forest products industry directly and indirectly employed about 100,000 people and was the primary manufacturing activity in 245 provincial municipalities, according to the office of the Minister of Natural Resources and Wildlife (le Ministère des Ressources naturelles et de la Faune). In the same year, $2.8 billion in revenue derived directly from logging. But a crisis was looming for Québec's forest products industry, and in Canada generally, triggered by the collision of an increasingly strong Canadian dollar against its U.S. counterpart, a downturn in the U.S. housing market, a drop of 20 percent in the allowable harvest (the provincial government of Québec owns 89 percent of the province's forest), increasing fuel and energy costs, and shrinking demand for newsprint. In addition, there was a recommendation to further reduce the allowable cut by 22 percent, with the combined reduction of 42 percent adding greater uncertainty about raw material availability in the province's forest products industry.[69]

At the same time many of the remote areas in which the province's forest products industry operated were losing population to urban areas, resulting in current and projected labor shortages in the forest sector. The Québec government tried to help, designating over $1 billion over a four-year period toward market diversification, technology development, and worker education. But even this was not nearly enough. In summer 2006, a representative of the forest industry indicated that the government should invest $2 billion "if it is really serious about kick starting the industry."[70] Unfortunately, these initiatives and calls to action added fuel to the claims by the United States that Canada is unfairly subsidizing its forest products industry, thereby negatively impacting U.S. jobs.[71] Additional controversies plagued Québec's forest products industry and damaged its public image at home. The Colombe Report, an outcome of a provincially commissioned study, found that Québec's forests were being harvested at an unsustainable rate. A smattering of headlines from 2006–7 told a story of a Québec forest products industry in crisis months before the economic downturn of the 2008 recession had become fully realized:

> Crise forestière: 106 usines fermées et 8000 emplois perdus depuis avril 2005 (Forestry crisis: 106 mills closed and 8000 jobs lost since April 2005), *Le Soleil*, October 4, 2006.
>
> Industrie forestière: Près de 7000 mises à pied en 18 mois (The forest industry: Almost 7000 laid off in 18 months), *Le Journal de Québec*, October 7, 2006.
>
> Y. Laprade, L'avenir s'annonce sombre pour l'industrie forestière (The future looks bleak for the forest industry), *Le Journal de Québec*, October 14, 2006.
>
> E. Emond, 1027 travailleurs au chômage (1,027 workers unemployed), *Le Journal de Montréal*, March 31, 2007.
>
> 12,000 emplois perdus depuis le début de la crise (12,000 jobs lost since the beginning of the crisis), *Le Journal de Québec*, March 31, 2007.

The future for Québec's logging workforce appeared bleak. Looking for opportunity, young people had been leaving the rural areas of the province for the cities since at least the time of the *la révolution tranquille* in the 1960s, and for those who remained, according to one observer, working in *la forêt intéresse peu les jeunes* (the forest holds little interest for the young).[72] The collision of forestry sector crises left the impression that there are no longer viable employment opportunities in Québec's logging woods, with one Québec newspaper headline asking, "La foresterie, un bon choix de carrière?" (Forestry, a good career choice?).[73] Others continued to look south to Maine for logging employment, exacerbating already existing tensions in the state's north woods.

In the meantime, and further accentuating the challenges confronting the future of Québec's logging workforce, according to a 2004 study, Québec-resident loggers who worked in Maine's border counties with that province (Aroostook, Franklin, Oxford, Piscataquis [located less than fifteen miles from Québec], and Somerset Counties) were significantly older (forty-nine years) and less educated (nine years) than their northern Maine counterparts (forty-four years and twelve years, respectively). Québec-resident loggers were also more experienced woods workers (thirty years versus twenty-two years) and traveled over twice as far (eighty-two miles versus thirty-six miles) to their logging woods workplaces than loggers from northern Maine.[74]

The relatively low level of formal education attained by Québec resident loggers in the 2004 study is not surprising. Loggers in eastern Canada have historically had the lowest education levels of major Canadian census groups, with between 79 and 87 percent of loggers in the eastern provinces of New Brunswick, Newfoundland, Ontario, and Québec possessing an elementary education or less, according to a 1966 monograph.[75]

The apparent lack of both occupational prestige and job satisfaction experienced by Québécois loggers described by the Public Affairs Research Center in 1968 was largely reaffirmed by a 2004 cross-cultural study of the logging community of eastern Québec. The authors found significant differences between Maine- and Québec-resident loggers in their responses to questions related to why they became loggers.[76] For example, Maine-resident loggers in the study region more often indicated that in choosing to become loggers, they felt less constrained by their education or the prospects of other employment than did loggers from Québec, many of whom appeared to be more resigned to logging, often suggesting that it was the only work that was available or that they knew how to do. In addition, significantly more Québec-resident loggers expressed strong intergenerational attachment to logging and indicated that they are loggers because that is what they have always done. However, despite this intergenerational attachment to their profession, far fewer Québécois loggers than those from Maine's north woods region would recommend logging to their offspring. What all of this—an aging and poorly educated Québécois logging workforce, a breakdown of familial attachment to logging, long weekly commutes to remote logging camps, the migration of youth from rural to urban Québec—portends for Maine's logging woods has yet to be fully realized.

"INDUSTRY OF THE PAST"?

Attempts to attract new industry to the state's forest products sector have been arrhythmic and the messaging somewhat tired, lacking a more business-friendly tone and substance. For example, while it is often repeated that Maine is 89 percent forested and that forests are a renewable resource, usually in efforts to convince others of the significance or potential of the state's forestry sector, one wonders why these attributes might be considered compelling reasons to maintain or grow a forestry business in Maine specifically. The 2005 Governor's Council appeared to agree. "Having one of the world's best supplies (quantity and quality) of fiber is no longer enough to insure the success or in fact the very continuation of Maine's forest products industry," it maintained.[77] Or as stated more bluntly by one Maine forestry opinion leader a decade or so later, "vague sentimental mush is not needed," when taking on the challenges facing the state's forestry sector. "Some advantages important in the past have now vanished. The world does not owe us jobs, tax revenues and prosperity because of our large stock of biomass." While Maine and Mainers may think that they need the forest industry, "the government cannot make the forest industry need Maine."[78] As a result, much of the nation's forestry production capacity has moved to more favorable forest growth and economic conditions in the southern United States and beyond.

Indeed, along with a more favorable climate and better growing conditions—not to mention the potential for year-round logging uninterrupted by a months-long mud season—the state of Georgia, for example, has more timberland than the states of Maine and neighboring New

Hampshire combined, as well as a more attractive business climate than does Maine. Importantly, Maine is not generally considered a business-friendly place, primarily owing to a comparatively restrictive regulatory environment and unfavorable corporate tax structure. "Maine lags the rest of New England and the nation because its taxes are so high no sensible person would ever invest anything in Maine," according to a 2006 Brookings Institution report.[79] While only partially explaining the continued downturn in the state's forestry sector, business climate is an area over which the state has some leverage, given a political will to change.

More recently, advocates for Maine's forest industry also point to the banality that forest growth is far outpacing forest removals in the state, failing to acknowledge that this is more likely a result of a struggling forest products sector—and therefore of less forest harvesting—than of vastly improved statewide forest stewardship. In this context, the spot fires ignited by a litany of controversies and conflicts between mill owners and loggers, between American loggers and their Canadian counterparts, and among various forest user and interest groups have served as a distraction to the greater challenge of creating a space where they can all coexist in a mutualistic relationship and a more innovative, diverse, and sustainable forest products business climate. Unfortunately, as we've seen, Maine's post–World War II forest industry has rarely appeared to be capable of developing lasting partnerships between loggers and mills—suppliers and consumers of the same raw material—despite seemingly endless studies of the challenges faced by the state's forestry sector, too often conducted by an "industry studying itself."

Despite its indispensable role as suppliers of wood raw material to the state's forest products manufacturing sector, Maine's logging community has at times been inexplicably excluded from discussions about the industry's future. For example, the elaborately titled 2005 document *Maine Future Forest Economy Project: Current Conditions and Factors Influencing the Future of Maine's Forest Products Industry* missed a significant opportunity to include the logging industry in its discussion of the state's forest economy.[80] An "initiative of the Department of Conservation—Maine Forest Service," the 474-page report makes no mention of the logging community as an integral and necessary dimension of the forest products industry, while devoting entire sections to everything from the state's forest inventory to its business climate and the transportation of its forest products. Such an oversight only served to further alienate Maine's loggers, leaving an impression that logging will take care of itself while the state focused its attention on forestland owners and wood- and fiber-consuming mills as the only important elements of the forest products supply chain. In addition, prior to 2000, there had been little research related to the state's logging community. Indeed, a comprehensive study of the logging community of the northern forest, conducted in 2000 by the University of Maine and funded by the U.S. Forest Service, which included mail surveys to over three thousand loggers in Maine, New Hampshire, New York, and Vermont, was met with ambivalence by Maine's state forestry office, whereas the region's other state forestry entities enthusiastically supported the project. The study went on to become a significant marker for understanding some of the social and economic challenges facing Maine's logging community, with its research approaches subsequently adopted in studies of logging communities in other states and in Québec.[81]

One example of the lack of cooperation between loggers and wood-consuming mills that ultimately burdens the entire forest products industry supply chain is the recurring problem of idle logging capacity. According to a University of Maine master's thesis, idle logging capacity is a result, at least in part, of wood-consuming mills purchasing raw material from a large number of independent logging contractors, thereby driving down mill-delivered wood prices and, in the worst case, intentionally playing one logger against another to lower the wood costs of mills.[82] Overall, this behavior resulted in a very inefficient wood supply system, with costs ultimately borne by all participants in that system, from landowner to logger to consuming mill. A 2004 forest industry-sponsored study found that the primary causes of unused logging capacity in Maine and several southern states were related to markets, restrictive quotas, weather and road conditions, and poor planning. The authors concluded that the wood supply system as a whole bears the cost of idle logging capacity, preventing it from competing globally. Participants in the wood supply system—mills, wood dealers, and loggers—should be able to manage these costs, the authors contended, but it would require a cooperative effort to make it happen. "Simply reducing the rates paid for logging services and expecting the market to adjust unassisted is wishful thinking and arguably counterproductive," the authors concluded. "Many of the causes of unused capacity appear to be within the control of receiving mills, wood dealers, and logging firms in the wood supply system. Losses due to market causes, primarily quotas, topped the list. Control of these will fall primarily on mills and dealers. Lost production due to mechanical and labor factors are issues for the logger to address," suggesting a need for the industry to cooperate throughout all elements comprising the forest products supply chain.[83]

While likely not a comfort to Maine's forest products industry, other aspects of the state's rural economy have also been in decline. Maine's agriculture sector, for example, has been facing similar challenges to its statewide relevance, although forests occupy a much greater percentage of the state's landscape. According to the 2017 Census of Agriculture, less than six percent of Maine's total land area was in agriculture, and the state's total farm acreage had declined in the five years since the 2012 census. The total number of farms in Maine fell by 7 percent during the same period, with the overall net cash farm income down 22 percent. Crops such as potatoes and blueberries may be locally important to some northern and Down East communities, but given Maine's generally marginal soil quality and unfavorable climate, agriculture is far from a game changer in the state's overall economy, and as census data indicate, its residual relevance appears to be fading quickly.[84] The same may be said of aspects of the state's forest industry, given significant losses in the pulp, paper, and biomass sectors. Yet one of Maine's seven legislatively designated strategic technology areas is "advanced technologies for forestry and agriculture," suggesting a state that appears to be out of sync with its own economic realities. That the state of Maine is generally ranked near the bottom regarding investments in research, including to its universities, further challenges a path forward for growing an innovative, data-informed, and more stable forestry sector benefiting from "advanced technologies" in wood utilization.[85]

To many advocates of a more diversified, forward-looking Maine economy, it is surprising that the state's forestry sector continues to function, despite data that show a consistent downward trajectory in forestry's contribution to Maine's employment and gross domestic product. A 2006 Brookings Institution report asserted that the state's "economy is struggling, with mixed success, to make the transition from a low labor-cost, resource-dependent industrial economy concentrated mainly in the rural areas to a high skill, innovation-driven post-industrial economy centered in urban regions." There has been a steady decline in Maine's manufacturing sector since 1970, including in forestry, where mechanization replaced people, with wages in natural resources at about

75 percent of the wages in this sector across the United States. "Maine's economic future lies in its cities," according to the report. "The long-term competitive advantage [of Maine's spruce-fir forest] is open to question,"[86] the overworked refrain that forests are renewable, that the state is 89 percent forested, and that its forest growth is exceeding harvest notwithstanding. As discussed, since the 2006 Brookings report matters have deteriorated considerably in the state's pulp, paper, and biomass sectors.

Yet the state has continued to prop up its forest products industry, as seen recently with subsidies to save Maine's waning biomass industry. One Maine governor decided to tether prospects for the state's postrecession economic recovery on what an observer referred to in 2016 as "industries of the past—timber harvesting and mining, for example," ostensibly as "part of his growth strategy, instead of focusing on growing, technology-driven enterprises. The Maine economy already has transitioned from one based on extracting natural resources to a largely urban service economy much like the rest of the U.S." When the governor took office, "he pledged to grow Maine's economy by strengthening the state's traditional industries." Quoting the governor during his 2012 State of the State address, "Through much of history, fishing, farming and forestry have been Maine's economic engine. We are committed to reviving these industries to get Maine working again." Since that time, the state's pulpwood and biomass industries have been in a state of accelerating crisis and contraction, the article's author citing a sharp decrease in paper-mill employment—from more than 15,000 in 1990 to 5,200 at the end of 2015—and the closure of five paper mills between 2011 and 2016. "Mill closures have had a ripple effect throughout the forest products industries," with loggers among those hardest hit by the closures. Instead of making decisions about Maine's future economy informed by a dispassionate consideration of the best available data, the governor increased timber harvesting on state lands and initiated yet another logger training program. "We keep trying to get the resource-based industries back into play, which isn't going to happen. We're not going to get new paper mills," the author quoted an University of Southern Maine economist as asserting. "Yet, our picture of the Maine economy among many if not most Mainers is still tied up in the natural resources sector."[87]

Contributing to the state's overall economic decline is the "curse of natural resources," a phenomenon typically describing the challenges facing some resource-rich developing countries that suffer from a lack of social, political, and economic infrastructure to either manage them sustainably or benefit from a value-added, resource-based economy. This paradox of plenty has been mentioned as contributing to the sorry state of Maine's forestry sector, as well its overall economic malaise, despite being surrounded by resource-abundant landscapes on land and on sea. Indeed, it has been postulated that in places suffering from this so-called poverty paradox there was less incentive for resource-dependent workers and their families to aspire to higher education levels and alternative skill development.[88] According to the U.S. Census Bureau, from 2015 to 2019, although Mainers graduated from high school at a rate of 92.6 percent—near the top among all states—only 31.8 percent earned a bachelor's degree or higher during the same period, the lowest percentage among all New England states.[89]

AN ENDLESS SWIRL

In the meantime, Maine's logging community has often found itself in the precarious, reactive position of chasing opportunities that pivot on the uncertainties of global politics and economics, persuaded to upgrade or expand already highly mechanized and capitalized logging sides in an attempt to meet the latest surge market demand, to be part of the next trend, the next political initiative. There are obvious risks for these logging business owners. As noted by an independent contractor who participated in a group meeting of seven loggers held in Gardner, Massachusetts, in 2005, "The [wood] markets may be uncertain, but equipment payments are the same every month." Nevertheless, many of Maine's independent loggers have continued to take the bait. Timing is everything, however, with logging businesses coming late to the game left holding expensive logging machinery but often faced with marginal logging chances, fading outlets for their harvested raw material, unfair wood procurement practices, suffocating regulations, or a lack of skilled manpower. Although some logging business owners have decided to be less reliant on the vagaries of wood markets and procurement processes by investing heavily in becoming more vertically integrated—owning everything from forestland to transportation and primary and secondary wood-processing facilities—such decisions require significant investment and assume even greater financial risk. Other logging businesses have diversified, with forest harvesting secondary in importance or complementary to earthmoving, trucking, and construction—sectors in Maine's economy that have demonstrated greater agility, resilience, and growth potential.

Unfortunately, Maine's overall economy has been relatively stagnant, and what growth has occurred has been taking place in the southeastern corner of the state, not in the state's more rural northern and western tiers. Rural Maine has been experiencing much more uncertain economic times. "[Rural] workers need good-paying jobs to replace the many losses in the pulp and paper industries," according to a 2017 Maine Center for Economic Policy release titled "A New Great Depression for Rural Maine."[90] Fishing, tourism, and forestry will not be enough to fuel Maine's future economy, asserted a *New York Times* report three years later.[91] The Maine Economic Development Strategy 2020–2029 placed the state "at an economic crossroads," acknowledging that Maine is one of the oldest and least diverse states in the nation. Jobs in Maine "don't pay enough," generally because of a weak value-added economy and the loss of jobs in manufacturing, a sector to which, during its heyday, the forest products industry had been the major contributor. Curiously, the report cited four thematic areas in which the state "has current strengths, there is growing global demand, and there is a potential for job creation," two of which were forest products and manufacturing—sectors that had been in a steep decline for years.[92]

There may have been a recalibration of this thinking, if Maine's 2020 Economic Recovery Committee was any indication. Convened by the governor to engage "economic experts and industry representatives from across Maine to develop specific policy recommendations to stabilize the state's economy and build a bridge to future prosperity in the wake of the COVID-19 pandemic," it comprised six subcommittees, including areas of demonstrable economic growth in the state, such as health care, hospitality, and education. Forestry and agriculture were also included but appeared to be lumped in the broader "manufacturing and natural resource-based industries" subcommittee.[93] With the dual objective of stabilizing and growing these six areas in the face of challenges presented by the prevailing pandemic, the initiative begs the question as to whether the governor's well-intentioned initiative could be an opportunity for a more decisive, critical, and forward-looking review of Maine's overall economic priorities based on the best available data, rather than on emotional and political attachments to the state's "traditional" industries. Put another way and borrowing from a former manager for Great Northern Paper Company writing in the *Bangor Daily News* in support of the establishment of a national park in the state's Katahdin region to bolster its economy in the wake of paper mill closures: "Nostalgia isn't an economic development plan."[94]

In the endless swirl of reports and plans, councils and commissions, stagnant economies and economic crossroads, and curious platitudes and stark realities, is there a promising future in Maine's logging woods and in its forestry sector, in an industry that has at times, over its long and colorful history, appeared to have gone haywire? An "industry of the past" looking for answers and a future, or perhaps simply biding its time, by studying itself, inexplicably expecting different, more durable outcomes? Whether one considers Maine's loggers to be victims or perpetrators of dysfunction in the state's forest products industry, participants in or simply observers of the various episodes comprising the forestry sector's history and its more recent state of affairs, is likely irrelevant. However, given its past, if Maine's working forests are going to recover and once again thrive, a less divisive, more mutualistic partnership that recognizes the value of all participants in the forest products supply chain is relevant.

None of this suggests that forestry, like agriculture, isn't locally important in parts of rural Maine and that it won't continue to contribute to a limited, sustainable rural economy and community, despite unfavorable statewide trends in both sectors that would indicate otherwise. Advances in wood utilization technology could offer the industry and the woods workers and rural communities that depend on it at least a short-term boost, maybe even the next innovation that, like the wood fiber pulp and paper revolution, would sustain the industry for generations.

Clearly, logging, like commercial fishing, has provided lasting historical and cultural identity to the state: Maine's north woods, farmer-lumberjacks, logging camps and river drives, *le bûcheron québécois*, choppers, swampers, teamsters and skidder operators, bonds and jobbers. But recalling the despondent northern forest logger who in 2005 declared that "tourists still want to see some of us around standing by the highway wearing plaid shirts," there's a catch to Maine's logging woods nostalgia. Indeed, when industry leaders claim in 2013 that the forestry sector "has a bright future" and that the state's forest and forestry sector are "thriving and remain an essential part of our state's economy," as that industry unraveled around them; when a governor, who was considered a friend of the forest industry, describes it in 2016 as "traditional"; and when woods workers in 2019 advocate for their profession by stating that they want to do "the work our fathers, grandfathers and great-grandfathers did before us," it may be time to listen more carefully to the Maine logging contractor who lamented: "I'm proud of what I do, but I'm getting to the point now where I don't tell people I'm a logger anymore; I say I'm a businessman in the forest products industry. I'm tired of telling everybody I'm a logger and they look at me like I have two heads."[95] And ask, *Why?*

NOTES

INTRODUCTION

1 Christopher Burns, "What Stands Out about LePage's Approach to the Economy: A Focus on Industries of the Past," *Bangor Daily News*, June 18, 2016. Burns's 2016 article cited Maine's economy as one of the slowest growing in the country, while Governor Paul LePage focused his energy on trying to revive Maine's manufacturing and natural resource sectors. Suggesting that the governor "relies heavily on the industries of the past—timber harvesting and mining, for example," the article decried the lack of attention being paid to growing "technology-driven enterprises." According to Burns, "When LePage took office, he pledged to grow Maine's economy by strengthening the state's traditional industries."

2 Although the regions are similar, in *Haywire* I use *northern forest* in the lowercase so as not to confuse it with the political and, at times, politicized Northern Forest region that was the subject of some controversy in the 1980s over concerns about regional land use management of the multistate region's mosaic of public and private land. The northern forest of this book includes northern New York, New Hampshire, Vermont, and Maine, as well as nearby Québec and New Brunswick. The states comprising the region are among the most heavily forested in the country: Maine is the most heavily forested state in the country (90 percent of the state's total land area is forested), and the land areas of New Hampshire (84 percent), and Vermont (78 percent) are predominantly forested. The Adirondack Park of northern New York, about the size of the state of Vermont, is 87 percent forested. *Haywire*'s northern forest is not a distinct political or ecological region but rather an interconnected place that shares a logging woods with generally similar history, culture, forest types, and forestry practices.

3 Quotes are from focused discussions and interviews of loggers conducted by the author while on the forestry faculties at the University of Maine and Laval University. The quotation from the Vermont logger was cited in Andrew Egan and Deryth Taggart, "Who Will Log? Occupational Choice and Prestige in Northern New England," *Journal of Forestry* 102, no. 1 (January/February 2004): 22.

CHAPTER ONE: "ALMOST NECESSARILY POOR"

1 Vernon H. Jensen, *Lumber and Labor* (New York: Farrar and Rinehart, 1945), 33.

2 Occupational prestige differs from occupational status, the latter referring to objective socioeconomic conditions associated with holding an occupation and the former describing more subjective evaluations of the social standing of an occupation. See Elliot A. Krause, *The Sociology of Occupation* (Boston: Little, Brown, 1971), and Charles B. Nam and Mary G. Powers, *The Socioeconomic Approach to Status Measurement* (Houston, TX: Cap and Gown, 1983). Albeno Garbin and Frederick Bates, in "Occupational Prestige: An Empirical Study of Its Correlates," *Social Forces* 40 (1961), described the public's perception of the prestige of various occupations as being rooted in several relatively subjective attributes, including the degree of interesting and challenging work, intelligence required to perform primary work functions, and the intrinsic nature of the work. In addition, Donald

Treiman, in *Occupational Prestige in Comparative Perspective* (New York: Academic Press, 1977), suggested that nonmanual workers generally have higher prestige than do manual workers. In contrast, occupational status is measured by the more objective elements of education and income, often derived from census data (Ronald M. Pavalko, *Sociology of Occupations and Professions* [Itasca, IL: F. E. Peacock, 1971]).

3 The term "lumberjack" has not been commonly used to describe woods workers of the northern forest region. Early on, the region's woods workers were most often referred to as "lumberers," "lumbermen," "shantymen," and "woodsmen." These terms eventually gave way to the now common and generic "logger" and to terms that were used to describe woods workers who performed specific logging-related work functions, for example, "chopper" for the woods worker who fells trees. Some logging-related work functions no longer exist in the northern forest. "Rafters" or "raftmen," for instance, were considered specialists who worked during spring log drives.

The term "farmer-lumberman" appears in C. Ross Silversides's pivotal 1997 *Broadaxe to Flying Shear*, about the evolution of logging mechanization and labor on the northern forest (C. Ross Silversides, *Broadaxe to Flying Shear: The Mechanization of Forest Harvesting East of the Rockies* [Ottawa: National Museum of Science and Technology, 1997]). More recently, in his book *Shredding Paper*, about the rise and fall of Maine's paper industry, Michael Hillard uses "farmer-logger" to describe participants in Maine's early agri-forestry economy (Michael G. Hillard, *Shredding Paper: The Rise and Fall of Maine's Mighty Paper Industry* [Ithaca, NY: ILR Press, 2020]).

For "occupational pluralists," see Silversides, *Broadaxe to Flying Shear*.

4 Ian Radforth, "The Shantyman," in *Labouring Lives: Work and Workers in Nineteenth-Century Ontario*, ed. Paul Craven (Toronto: Toronto University Press, 1995), 214.

5 Donald MacKay, "The Canadian Logging Frontier," *Journal of Forest History* 23, no. 1 (January 1979): 4–17.

6 Even today, the best conditions for logging in the northern forest are during periods of winter-hardened and snow-protected skid trails and log landings—staging areas for skidded logs, rubber-tired skidders, feller-bunchers and logging trucks. Spring breakup, with its warming temperatures, seasonal rains, and muddy trails, signals a break in the logging season for most loggers in the region, who resume woods work only when drier and warmer late spring and summer conditions return. While mechanization has lengthened the logging year somewhat, changes in the region's climate, such as earlier "spring breakup" and therefore an earlier "mud season," have challenged this.

7 James Wesley McNutt, "Labor Relations in the Forest," Weyerhaeuser Lecture Series, Faculty of Forestry and Landscape Architecture, University of Toronto, November 1978, 14.

8 Andrew Egan, *Adirondack Hard Times: Evolution of a Rich Man's Paradise* (Charleston, SC: History Press, 2021).

9 Graeme Wynn, "Deplorably Dark and Demoralized Lumberers? Rhetoric and Reality in Early Nineteenth-Century New Brunswick," *Journal of Forest History* 24, no. 4 (October 1980): 168–87.

10 Richard W. Judd, "Lumbering and the Farming Frontier in Aroostook County, Maine, 1840–1880," *Journal of Forest History* 28, no. 2 (April 1984): 56–67, quotations on 57, 60.
11 Robert Pike, *Tall Trees, Tough Men* (New York: W. W. Norton, 1967).
12 Béatrice Craig, "Agriculture and the Lumberman's Frontier in the Upper St. John Valley, 1800–70," *Journal of Forest History* 32, no. 3 (July 1988): 125–37 (quotation on 125).
13 Judd, "Lumbering and the Farming Frontier," 56.
14 David Vail, "How to Tell the Forest from the Trees," *Technology in Society* 11 (1989): 348.
15 Félix-Antoine Savard, *Menaud, maître-draveur* (Québec: Bibliotheque Québécoise, 1992), 37. Author's translation, but with influences from the 1947 translation of Alan Sullivan, the Ryerson Press, Toronto.
16 Jeremy Belknap, *The History of New Hampshire (1784–92)*, vol. 3 (Boston, 1813).
17 Craig, "Agriculture and the Lumberman's Frontier," 125, 136.
18 Judd, "Lumbering and the Farming Frontier," 56–67 (quotation on 65).
19 Wynn, "Deplorably Dark and Demoralized Lumberers?," 168–87 (quotation on 172).
20 Camille Legendre, *Le travailleur forestier Québécois* (Québec: Presse de l'Université du Québec, 2005), 158. Author's translation.
21 Burt Kirkland, "Effects of Destructive Lumbering on Labor," *Journal of Forestry* 18, no. 4 (April 1920): 318.
22 Egan, *Adirondack Hard Times.*
23 Wynn, "Deplorably Dark and Demoralized Lumberers?," 186–87.
24 Egan, *Adirondack Hard Times.*
25 Alfred Donaldson, *A History of the Adirondacks*, 2 vols. (New York: Century, 1921), 2:153.
26 *Third Annual Report of the Commissioners of Fisheries, Game and Forests of the State of New York* (Albany, NY: Wynkoop, Hallenbeck, Crawford, 1898), 350.
27 *Fourth Annual Report of the Commissioners of Fisheries, Game and Forests of the State of New York* (Albany, NY: Wynkoop, Hallenbeck, and Crawford, 1899), 330.
28 Pike, *Tall Trees, Tough Men*, 54.
29 Edward Augustus Kendall, *Travels through the Northern Parts of the United States, in the Years 1807 and 1808*, vol. 3 (New York: I. Riley, 1809), 89–93.
30 MacKay, "The Canadian Logging Frontier," 8.
31 Richard Wood, *A History of Lumbering in Maine: 1820–1861* (Orono: University of Maine Press, 1935), 185.
32 Timothy Dwight, *Travels in New England and New York, vol. 2* (New Haven, CT, 1821), 527.
33 John McGregor, *Historical and Descriptive Sketches of the Maritime Colonies of British America* (London, 1828), 266.
34 McNutt, "Labor Relations in the Forest," 40.
35 Jensen, *Lumber and Labor*, 314.
36 Wayne E. Reilly, "Hundreds of Immigrant Loggers Stranded in Bangor a Century Ago," *Bangor Daily News*, April 27, 2014, https://bangordailynews.com/2014/04/27/living/hundreds-of-immigrant-loggers-stranded-in-bangor-a-century-ago/.

Cant dogs and setting poles are manual wood-handling tools used in logrolling and river driving, respectively.

37 Kirkland, "Effects of Destructive Lumbering on Labor," 318.

38 Wynn, "Deplorably Dark and Demoralized Lumberers?," 168, 174.

39 Frederick E. Olmsted, "Business Phases of Forest Devastation," *Journal of Forestry* 18, no. 4 (April 1920): 312, https://academic.oup.com/jof/article-abstract/18/4/311/4752063?redirectedFrom=fulltext). A "blanket stiff" is defined in *Webster's* as "an itinerant, usually unskilled, laborer who travels with a blanket roll." See Merriam-Webster.com, s.v. "blanket stiff," https://www.merriam-webster.com/dictionary/blanket%20stiff.

40 Wilson Compton, "Recent Developments in the Lumber Industry," *Journal of Forestry* 30, no. 4 (April 1932): 442.

41 Pike, *Tall Trees, Tough Men*, 275.

42 Pike, *Tall Trees, Tough Men*, 275.

43 *Webster's* defines *"jobber"* as "a person who works by the job". *Merriam-Webster.com, s.v.* "jobber," https://www.merriam-webster.com/dictionary/jobber.

44 McNutt, "Labor Relations in the Forest," 14.

45 Jensen, *Lumber and Labor*, 3, 4.

CHAPTER TWO: "THE BEST WORKERS THEY HAD"

1 Vernon H. Jensen, *Lumber and Labor* (New York: Farrar and Rinehart, 1945), 33; David Vail, "The Internal Conflict: Contract Logging, Chainsaws, and Clearcuts in Maine Forestry," in *Who Will Save the Forests? Knowledge, Power and Environmental Destruction*, ed. Tariq Banuri and Frédérique Apffel-Marglin (London: Zed Books, 1993), 142–89.

2 Richard Donovan and Elizabeth Swain, *Maine Woods Labor Supply, 1984–85* (Augusta: Maine Department of Conservation, 1986).

3 James Wesley McNutt, "Labor Relations in the Forest," Weyerhaeuser Lecture Series, Faculty of Forestry and Landscape Architecture, University of Toronto, November 1978; Bill Parenteau, "The Rise of the Small Contractor: A Study of Technological and Structural Change in the Maine Pulpwood Industry" (master's thesis, University of Maine at Orono, 1986).

4 In *Haywire* I use the word *nativism* as a "policy of favoring native inhabitants as opposed to immigrants," as defined by *Webster's*. Merriam-Webster.com, s.v. "nativism," https://www.merriam-webster.com/dictionary/nativism.

5 Louis Hémon, *Maria Chapdelaine* (Montreal: Bibliothèque Québécoise, 1990), 67, 68. Author's translation.

6 Alfred Donaldson, *A History of the Adirondacks*, 2 vols. (New York: Century, 1921), 2:151; R. L. Marston, "The Lumbering Industry in Maine" (master's thesis, Yale University, 1902), 81.

7 Kerri Arsenault, *Mill Town: Reckoning with What Remains* (New York: St. Martin's, 2020).

8 According to Andrew Birden, "Lawmakers in 1919 Tried to Neuter Acadians by Banning Their Language in Schools," *Bangor Daily News*, March 1, 2019, https://bangordailynews.com/2019/03/01/news/lawmakers-in-1919-tried-to-neuter-acadians-by-banning-their-language-in-schools/: "On April 1, 1919, the Maine Legislature passed an English-only bill ["Chapter 146, Acts and Resolves as Passed

by the Seventy-Ninth Legislature of the State of Maine, 1919," http://lldc.mainelegislature.org/Open/LegRec/_79/Joint/LegRec079_01_p0001-0086_Index_R1.pdf] that targeted the Acadians of Aroostook County [Maine's northernmost county]. It required that schools educate children exclusively using the English language. In the Crown of Maine, this gave rise to the oft bitterly repeated phrase, 'I will not speak French in school,' which originated from teachers punishing children for speaking casual French in classrooms by forcing the child to write the phrase over and over again."

9 According to the *Maine Law Review*, Chapter 7 of Public Laws of Maine for 1907 read: "Whoever enters into an agreement to labor for another in any lumbering operation and in consideration thereof receives any advance in goods, money or transportation, and unreasonably and with intent to defraud, fails to enter into said employment as agreed, and to labor for a sufficient length of time to reimburse his employer for said advances and expenses of transportation, shall be punished by a fine of not exceeding $10 or by imprisonment not exceeding 30 days." See *Maine Law Review* 4, no. 7 (May 1911): 246.

10 Pam Belluck, "Long-Scorned in Maine, French Has a Renaissance." *New York Times*, June 4, 2006; "K.K.K. and I.W.W. Wage Drawn Battle in Greenville," *Portland Press Herald*, February 5, 1924; Michael Hillard, *Shredding Paper: The Rise and Fall of Maine's Mighty Paper Industry* (Ithaca, NY: ILR Press, 2020).

11 "The Ku Klux Klan Fights the Battle of Greenville, Maine," New England Historical Society, updated 2021, https://www.newenglandhistoricalsociety.com/the-ku-klux-klan-fights-the-battle-of-greenville-maine/.

12 "The Ku Klux Klan Fights."

13 Arsenault, *Mill Town*, 51.

14 Mark Paul Richard, *Not a Catholic Nation: The Ku Klux Klan Confronts New England in the 1920s* (Amherst: University of Massachusetts Press, 2015), 42.

15 "K.K.K. and I.W.W. Wage Drawn Battle."

16 Richard, *Not a Catholic Nation*. It was suggested by Richard Rajala in C. Ross Silversides's 1997 *Broadaxe to Flying Shear* that among the challenges to organizing logging labor in Ontario during the early 1900s were "the isolated location of camps, the seasonal nature of logging, ethnic diversity, and the tendency of men to view themselves as temporary wage-earners or farmers." See C. Ross Silversides, *Broadaxe to Flying Shear: The Mechanization of Forest Harvesting East of the Rockies* (Ottawa: National Museum of Science and Technology, 1997).

17 Ferdinand Silcox, "Forestry and Labor," *Journal of Forestry* 18, no. 4 (1920): 317.

18 Richard, *Not a Catholic Nation*.

19 Camille Legendre, *Le travailleur forestier Québécois* (Québec: Presse de l'Université du Québec, 2005), 4. Author's translation.

20 *La révolution tranquille* represented a significant cultural shift in the province of Québec, from a conservative, rural agrarian society controlled by a paternalistic Catholic church to a more nonsectarian, progressive society that valued innovation and liberalism.

21 Gérald Fortin and Émile Gosselin, "La professionnalisation du travail en forêt," *Recherches Sociographiques* 1, no. 1 (January–March 1960): 33–60.

22 Public Affairs Research Center, *Study of Problems Relative to Obtaining a Continuing Supply of Domestic Workers for Wood Operations in Maine* (Brunswick, ME: Bowdoin College, 1968).

23 J. Carter, "Woods Labor Review in Maine," *Northeastern Logger*, June 1960, 18–19, 46.

24 Robert S. Bond and Charles H. Wolfe, "Status and Outlook of Bonded Canadian Woodworkers in Maine," *Northern Logger and Timber Processor*, May 1978, 6–7, 44, 47–48 (quotation on 7).

25 Vail, "Internal Conflict," 185.

26 Legendre, *Le Travailleur Forestier Québécois*, 397. Author's translation.

27 Bond and Wolfe, "Bonded Canadian Woodworkers in Maine," 6.

28 Leo Thibodeau, "Effective Recruiting and Training of Woods Workers and Loggers," *Northern Logger and Timber Processor*, May 1965, 12–13, 31, 35 (quotation on 12).

29 Thibodeau, "Effective Recruiting and Training," 12.

30 Public Affairs Research Center, *Study of Problems*; Vail, "Internal Conflict," 142–89.

31 Vail, "Internal Conflict," 158–59.

32 Bond and Wolfe, "Bonded Canadian Woodworkers in Maine," 6–7, 44, 47–48.

33 Philip Cottell, *Occupational Choice and Employment Stability among Forest Workers*, Yale University School of Forestry and Environmental Studies Bulletin No. 82 (Vancouver: Pulp and Paper Research Institute of Canada, 1974), 161.

34 Duncan Campbell and Edward Power, *Manpower Implications of Prospective Technological Changes in the Eastern Canadian Pulpwood Logging Industry* (Ottawa: Research Branch, Program Development Service, Department of Manpower and Immigration, Canada, 1966).

35 McNutt. "Labor Relations in the Forest," 40.

36 Robert S. Bond, *A Report on Bonded Canadian Labor in New England's Logging Industry* (Amherst: Massachusetts Agricultural Experiment Station, 1977).

37 *Cut and Run*, directed by Richard Searls (Bucksport, ME: Northeast Historic Film, 1980), DVD.

38 Statistics Canada, cite in Andrew Egan and Deryth Taggart, "Who Will Log in Maine's North Woods? A Cross-Cultural Study of Occupational Choice and Prestige," *Northern Journal of Applied Forestry* 21, no. 4 (2004): 202.

39 Maine Department of Labor, cited in Egan and Taggart, "Who Will Log in Maine's North Woods?," 202.

40 Parenteau, "Rise of the Small Contractor"; Vail, "Internal Conflict," 159.

41 Roger Ferragne, "1973 Québec Study on Woodlands Labor Shortage," in *Manpower—Forest Industry's Key Resource*, ed. Lloyd C. Irland, Yale University School of Forestry and Environmental Studies Bulletin No. 86, papers presented at the 39th Industrial Forestry Seminar, New Haven, CT, May 20–24, 1974 (New Haven, CT: Yale University, 1975), 158–70, data found on 158.

CHAPTER THREE: TOO CLOSE TO THE STUMP

1 Beth Braverman, "The 10 Most Dangerous Jobs in America," CSNBC.com, December 28, 2019, https://www.cnbc.com/2019/12/27/the-10-most-dangerous-jobs-in-america-according-to-bls-data.html.

2 Deaths in the Woodlands Parish, *Northeastern Logger*, November 1953.

3 Deaths in the Woodlands Parish, *Northeastern Logger*, February 1954.

4 "The Work: Good Pay, Good Work, Good Union," B.C. Forest Discovery Centre (2004), www.camptocommunity.ca/english/work/unions.html. Forest According to *Webster's*, a *buckwheater* is "a green or inexperienced logger." See Merriam-Webster.com, s.v. "buckwheater," https://www.merriam-webster.com/dictionary/buckwheater.

5 William F. Fox, *A History of the Lumber Industry in the State of New York*, United States Department of Agriculture, Bureau of Forestry, Bulletin 34 (Washington, D.C.: Government Printing Office, 1902),16.

6 Fox, *History of the Lumber Industry*, 16; Nick Sambides, "Logger Pinned to Tree Cuts Himself Free with Chainsaw," *Bangor Daily News*, August 20, 2020.

7 *Bangor Register*, February 13, 1828.

8 James Wesley McNutt, "*Labor Relations in the Forest*," Weyerhaeuser Lecture Series, Faculty of Forestry and Landscape Architecture, University of Toronto, November 1978, 40.

9 Jennifer L. Bell and Shawn T. Grushecky, "Evaluating the Effectiveness of a Logger Safety Training Program," *Journal of Safety Research* 37, no. 1 (2006): 53–61.

10 *Cut and Run*, directed by Richard Searls (Bucksport, ME: Northeast Historic Film, 1980), DVD.

11 Richard Donovan and Elizabeth Swain, *Maine Woods Labor Study, 1984–85* (Augusta: Maine Department of Conservation, 1986), 9.

12 McNutt, "Labor Relations in the Forest," 6.

13 Searls, *Cut and Run*. The project director for Searls's documentary was University of Maine history professor Howard Schonberger, described by one source as "an activist for democratic socialism." Having passed away in 1991, the University of Maine continued to honor him through its Howard B. Schonberger Peace and Social Justice Memorial Lecture Series. For example, see https://mainecampus.com/2019/03/lecture-highlights-lesser-known-stories-of-martin-luther-king-jr-s-life/.

14 Quotations from meetings between New Hampshire and Maine loggers with author.

15 A misery whip is a two-man crosscut saw.

16 "Stumpage" is defined as (a) the value of standing timber and (b) uncut marketable timber. See *Merriam-Webster's Collegiate Dictionary*, 11th ed. (2003), s.v. "stumpage."

17 Andrew Egan and Deryth Taggart, "Who Will Log? Occupational Choice and Prestige in Northern New England," *Journal of Forestry* 102, no. 1 (2004): 22; Andrew Egan, "Characteristics of New York's Logging Businesses and Logging Business Owners," *Northern Journal of Applied Forestry* 26, no. 3 (2009): 108.

18 David Vail, "The Internal Conflict: Contract Logging, Chainsaws, and Clearcuts in Maine Forestry," in *Who Will Save the Forests? Knowledge, Power and Environmental Destruction*, ed. Tariq Banuri and Frederique Apffel-Marglin (London: Zed Books, 1993), 157–58.

19 David Vail, "How to Tell the Forest from the Trees," *Technology in Society* 11 (1989): 361.

20 Egan and Taggart, "Who Will Log?, 20–25.

21 Andrew Egan and Deryth Taggart, "Public Perception of the Logging Profession in Maine and Implications for Logger Recruitment," *Northern Journal of Applied Forestry* 26, no. 3 (2009): 93–98.

22 Egan and Taggart, "Who Will Log?," 20–25; Bennet H. Leon and Jeffrey G. Benjamin. "A Survey of Business Attributes, Harvest Capacity and Equipment Infrastructure of Logging Businesses in the Northern Forest," University of Maine School of Forest Resources, February 22, 2012, https://maineforest.org/wp-content/uploads/2013/03/Survey-on-logging-businesses.pdf, 26.

23 Leon and Benjamin, "A Survey of Business Attributes," https://maineforest.org/wp-content/uploads/2013/03/Survey-on-logging-businesses.pdf, 13.

24 Deryth Taggart and Andrew Egan, "Occupational Choice among Loggers in Maine's Northern Forest and Southern Counties," *Adirondack Journal of Environmental Studies* 9, no. 2 (Fall/Winter 2002): 31–36 (data on 33).

25 See "Preventing Injuries and Deaths of Loggers," DHHS (NIOSH) Publication No. 95–101, May 1995, https://www.cdc.gov/niosh/docs/95-101/default.html; John R. Myers and David E. Fosbroke, "Logging Fatalities in the United States by Region, Cause of Death, and Other Factors—1980 through 1988," *Journal of Safety Research* 25, no. 2 (Summer 1994): 97–105; and Andrew Egan, "Hazards in the Logging Woods—Who's Responsible?," *Journal of Forestry* 94, no. 7 (July 1996).

During the same period, I earned $4.50 an hour felling timber—"chopping"—for a logging company in central New Hampshire. According to the U.S. Department of Labor, the minimum wage in New Hampshire in 1980 was $3.10 per hour, the same as the federal minimum wage. Importantly, neither I nor most logging employees in the region received traditional employment benefits, including health care. Like loggers before and since, woods workers logged because they enjoyed the work, the woods, and the risk.

26 United States Bureau of Labor Statistics, *Injuries in the Logging Industry: Bulletin of the United States Bureau of Labor Statistics, No. 2203* (Washington, D.C.: U.S. Department of Labor, Bureau of Labor Statistics, June 1984), 1–2, 12.

27 C. Ross Silversides, *Broadaxe to Flying Shear: The Mechanization of Forest Harvesting East of the Rockies* (Ottawa: National Museum of Science and Technology, 1997), 174.

28 The in-woods pathways over which harvested logs are dragged, or "skidded," originally by animal power, are called *skid trails* or *skid roads*. The latter term, coupled with the notion that loggers were often considered destitute and rowdy vagabonds, eventually gave rise to the term *skid row*. Dictionary.com defines "skid row" as a "squalid district inhabited by derelicts and vagrants; also, a life of impoverished dissipation. This expression originated in the lumber industry, where it signified a road or track made of logs laid crosswise over which logs were slid. Around 1900 the name Skid Road was used for the part of a town frequented by loggers, which had many bars and brothels, and by the 1930s the variant skid row, with its current meaning, came into use." During Bangor's mid- to late-nineteenth-century heyday as the world's largest lumber port, the city's Exchange Street, "widely known for its hotels, rooming houses, taverns, and brothels, not to mention logging supply houses," was a skid row. See Donald Wilson, *Logging and Lumbering in Maine* (Charleston, SC: Arcadia Publishing, 2001), 128. It was located in a rather amorphous area of the city known variously as Hell's Half Acre, Devil's Half Acre, or simply The Acre, "a place where loggers, sailors and other workingmen gathered to spend cash on whiskey and women. Many of these men were transients, often immigrants, who traveled with jobs and

the seasons." See Wayne Reilly, "Where Was the Devil's Half Acre?," *Bangor Daily News*, September 16, 2012. Much of Bangor's notorious district was destroyed during the Great Fire of 1911.

29 Andrew Egan, "Hazards in the Logging Woods—Who's Responsible?," *Journal of Forestry* 94, no. 7 (July 1996): 20.

30 William C. Osborn, *The Paper Plantation: Ralph Nader's Study Group Report on the Pulp and Paper Industry in Maine* (New York: Grossman Publishers, 1974), 158.

31 Egan, "Hazards in the Logging Woods," 20.

32 Andrew Egan, "The Introduction of a Comprehensive Logging Safety Standard in the USA," *Journal of Forest Engineering* 9, no. 1 (1997): 17–23; Andrew Egan and Carol Alerich, "'Danger Trees' in Central Appalachian Forests of the United States: An Assessment of Frequency of Occurrence," *Journal of Safety Research* 29, no. 2 (1998): 77–85.

33 John R. Myers and David E. Fosbroke, "The Occupational Safety and Health Administration Logging Standard: What It Means for Forest Managers," *Journal of Forestry* 93, no. 11 (November 1995): 35–36.

34 Egan, "Hazards in the Logging Woods," 19.

35 Andrew Egan, "Reducing Forest Road Erosion: Do Foresters and Logging Contracts Matter?," *Journal of Forestry* 97, no. 8 (1999): 36–39.

36 Andrew Egan and Stephanie Phillips, "Timber Marking as a Forest Operations Consideration," *Forest Products Journal* 56, no. 9 (2006): 65–70.

37 Bureau of Labor Statistics, *Injuries in the Logging Industry*.

38 Myers and Fosbroke, "Logging Fatalities in the United States," 99.

39 United Stated Department of Health and Human Services (1994), "Logging Industry Fatality Rate for 1989—In-house Records, cited in Myers and Fosbroke, "Occupational Safety and Health Administration Logging Standard," 34.

40 Andrew Egan, Curt Hassler, and Shawn Grushecky, "Are Logger Certification and Training Worthwhile? A View from West Virginia's Logging Community," *Forest Products Journal* 47, no. 7/8 (1997): 46–50.

41 Andrew Egan, Curt Hassler, and Sean Grushecky, "Logger Certification and Training: A View from West Virginia's Logging Community," *Forest Products Journal* 47, no. 7/8 (1997): 49.

42 Bell and Grushecky, "Logger Safety Training Program," 53–61.

43 Carin Sundstrom-Frisk, "Behavioural Control through Piece-Rate Wages," *Journal of Occupational Accidents* 6 (1984): 49.

44 Occupational Health and Safety Administration, 29 C.F.R. Parts 1910, Logging operations, and 1928 (October 12, 1994), 51699.

45 Egan and Alerich. "'Danger Trees' in Central Appalachian Forests," 77.

46 Silversides, *Broadaxe to Flying Shear*, 174.

47 Mark Hardison, "A Maine Woodcutter Talks about Safety in the Woods," Blaine House Conference on Forestry, Maine Department of Conservation, December 1984, 9.

48 Sundstrom-Frisk, "Behavioural Control through Piece-Rate Wages," 58.

49 Keith Mason, "The Effect of Piecework on Accident Rates in the Logging Industry," *Journal of Occupational Accidents* 1 (1977): 281–94.

50 Maine's "industrial forest" refers to the corporate ownership of much of the state's northern forest, originally by seven large forest products enterprises—Great Northern-Nekoosa, International Paper, Scott Paper, Diamond International, St.

Regis Paper, Georgia-Pacific, and Oxford Paper—according to William Osborn, primarily for the production of raw material for the pulp and paper industry. The state's industrial forests have at times been referred to pejoratively as a "paper plantation" and "chilly banana republic," following Osborn's 1974 book *The Paper Plantation*.

CHAPTER FOUR: CUT LOOSE

1 Peter MacDonald, and Michael Clow. "What a Difference a Skidder Makes: The Role of Technology in the Origins of the Industrialization of Tree Harvesting Systems," *History and Technology* 19, no. 2 (2003): 127–149 (quotations on 128).

2 A industrial forester is employed by a wood-consuming mill as a procurement or landowner assistance agent of the mill, the latter providing forest management advice and other forestry services to forestland owners, generally with the understanding, contractual or otherwise, that landowners would sell them their logs and/or stumpage when it came time to harvest. The author was briefly a landowner assistance forester for a sawmill in central New Hampshire until the mill burned in 1985.

3 Bill Parenteau, "The Rise of the Small Contractor: A Study of Technological and Structural Change in the Maine Pulpwood Industry" (master's thesis, University of Maine at Orono, 1986), unpaginated abstract.

4 Parenteau, "Rise of the Small Contractor."

5 Leo Thibodeau, "Effective Recruiting and Training of Woods Workers and Loggers," *Northern Logger and Timber Processor*, May 1965, 12.

6 George Carlisle, "Comments on Mr. Leo Thibodeau's Paper," *Northern Logger and Timber Processor*, May 1965, 13.

7 Robert S. Bond and Charles H. Wolfe, "Status and Outlook of Bonded Canadian Woodworkers in Maine," *Northern Logger and Timber Processor*, May 1978, 6–7, 44, 47–48.

8 Bond and Wolfe, "Bonded Canadian Woodworkers in Maine," 6.

9 David Vail, "The Internal Conflict: Contract Logging, Chainsaws, and Clearcuts in Maine Forestry," in *Who Will Save the Forests? Knowledge, Power and Environmental Destruction*, ed. Tariq Banuri and Frederique Apffel-Marglin (London: Zed Books, 1993), 142–89. p. 159, 185.

10 Thibodeau, "Effective Recruiting and Training," 12.

11 J. Carter, "Woods Labor Review in Maine," *Northeastern Logger*, June 1960, 18.

12 Robert S. Bond, *A Report on Bonded Canadian Labor in New England's Logging Industry* (Amherst: Massachusetts Agricultural Experiment Station, 1977).

13 Vail, "Internal Conflict," 152.

14 Dan Neumann, "Maine Loggers Fight to Unionize the North Woods," April 30, 2019, Mainebeacon.com, mainebeacon.com/maine-loggers-fight-to-unionize-the-north-woods/.

15 Quotation from logger participating in a focused discussion led by author, Vermont, 2005.

16 Robert Pike, *Tall Trees, Tough Men* (New York: W. W. Norton, 1967), 264.

17 David Vail, "How to Tell the Forest from the Trees," *Technology in Society* 11 (1989): 355.

18 Frank A. Reed, *Lumberjack Sky Pilot* (Lakemont, NY: North Country Books, 1976), 16.

19 S. Hall, "Problems of Contract Logging," *Northern Logger and Timber Processor*, September 1959, 12–13.

20 Phyllis Austin, "Hard Times in Irving's Woods," *Maine Times*, May 25–31, 2000.

21 Vail, "Internal Conflict," 152.

22 Hall, "Problems of Contract Logging," 12–13.

23 I spent some months marking timber along the Beebe River railroad grade while a forester for a New Hampshire forest management company. The tracks were pulled in 1942 by the Draper Corporation, which owned the logging railroad at the time.

24 See Hall, "Problems of Contract Logging," 13. It's unclear from his writing whether "purchase logs" represented what is often referred to in some wood procurement circles as "gatewood"—uncontracted logs that arrive at the mill without an agreement and often in the absence of a prescheduled or known mill delivery day or time. Gatewood is often assumed to be the least expensive wood that a mill can buy, because it typically does not entail much, if any, management or risk on the part of the consuming mill, with loggers assuming all costs and risks associated with moving wood from the stump to the mill. However, its delivery may also be unpredictable, making log inventory planning and sourcing a challenge. I used the term when I was a procurement forester for a sawmill in central New Hampshire, where the mill maintained the option of turning gatewood away, especially if the logs' specifications (e.g., lengths, scaling diameters, species, quality) were inconsistent with the mill's purchasing policies. In northern Maine, however, where industrial forestlands prevail, purchased wood generally referred to wood that was not from one's company lands, since the companies couldn't always provide 100 percent of their mill's wood requirements from their own lands.

One northern Maine industry forester once bumptiously suggested that I "drop the term *gatewood*" after using it in a conversation with him, as it wasn't part of Maine's north woods lexicon, according to the forester. However, the term was not uncommon throughout much of the rest of the northern forest region and across the United States. His comment highlighted differences in forestry and logging jargon and culture in the region, as well as the risks associated with assuming that there is a single, prevailing logging or logging woods culture, either in the northern forest region or in the state of Maine. Indeed, the term *landowner*, as used by Maine's industrial foresters, generally meant "industrial landowner" to many northern Maine foresters, whereas in other parts of the northern forest region the term may more likely have referred to nonindustrial forest owners, who owned most of that region's working forest.

25 Hall, 13.

26 Robert L. Moodie, "Contract Logging in the Pulpwood Business," *Northeastern Logger*, September 1959, 14–18 (quotation on 14).

27 Herbert Winer, "Methods of Paying Woods Workers," *Northeastern Logger*, September 1961, 34.

28 Michael Hillard, "Interview with Jon Falk," July 27, 2004, University of Southern Maine Digital Commons, https://digitalcommons.usm.maine.edu/strike/13/.

29 Board-foot log rules estimate the amount of board feet that could be sawn from logs of different diameters and lengths. However, the assumptions built into log

rules do not always accurately measure the volume of logs or reflect production processes extant at sawmills. Later on, weight scaling of sawlogs helped alleviate some of the conflict between logger and scaler, but weight scaling of sawlogs (as opposed to weight scaling of pulp and woody biomass raw material) is more common in the southern United States than in the northern forest and doesn't always adequately account for sawlog quality, or "grade."

30 Moodie, "Contract Logging in the Pulpwood Business," 15.

31 Moodie, 15.

32 Moodie, 18.

33 Richard Waldron, "Contract Logging," *Northeastern Logger*, September 1959, 19.

34 Austin, "Hard Times in Irving's Woods," 7, 4.

35 Hall, "Problems of Contract Logging," 13.

36 Matthew S. Carroll, *Community and the Northwestern Logger: Continuities and Changes in the Era of the Spotted Owl* (Boulder, CO: Westview Press, 1995), 106.

37 This isn't to say that very formative relationships between loggers and foresters did not exist in the northern forest. While logging in New Hampshire, the Bickford Logging crew that I chopped and skidded for worked constructively with Scott Paper forester Jack Wadsworth, who understood logging firsthand and looked at our relationship as a partnership that worked on behalf of the landowner and forest stewardship, rather than one that was supervisory, paternalistic, or adversarial.

38 Vail, "How to Tell the Forest from the Trees," 348.

39 Vail, "Internal Conflict," 144–45.

40 William C. Osborn, *The Paper Plantation: Ralph Nader's Study Group Report on the Pulp and Paper Industry in Maine* (New York: Grossman Publishers, 1974), 300.

41 Vail, "Internal Conflict," 142–89.

42 Egan and Taggart, "Who Will Log?," 20–25; Andrew Egan and Deryth Taggart, "Who Will Log in Maine's North Woods? A Cross-Cultural Study of Occupational Choice and Prestige," *Northern Journal of Applied Forestry* 21, no. 4 (2004): 200–208.

43 *Cut and Run*, directed by Richard Searls (Bucksport, ME: Northeast Historic Film, 1980), DVD.

44 Osborn, The Paper Plantation, 138–39.

45 L. P. Bissell, "Can the Woods Labor Problem Be Solved?" *Northern Logger and Timber Processor*, 1968, 16, 62.

CHAPTER FIVE: PUSHING BACK

1 William C. Osborn, *The Paper Plantation: Ralph Nader's Study Group Report on the Pulp and Paper Industry in Maine* (New York: Grossman Publishers, 1974), 165.

2 Michael Hillard, "Interview with Mel Ames," July 2004, University of Southern Maine Digital Commons, "Maine Woodman's Association Strike of 1975," digitalcommons.usm.maine.edu/strike/.

3 Michael Hillard, "Interview with Don Fontaine," July 15, 2004, University of Southern Maine Digital Commons, "Maine Woodman's Association Strike of 1975,"digitalcommons.usm.maine.edu/strike/.

4 Richard Donovan and Elizabeth Swain, *Maine Woods Labor Study, 1984–85* (Augusta: Maine Department of Conservation, 1986), 7–8.

5 Hillard, "Interview with Don Fontaine."

6 Robert S. Bond, *A Report on Bonded Canadian Labor in New England's Logging Industry* (Amherst: Massachusetts Agricultural Experiment Station, 1977), 11.
7 Osborn, *The Paper Plantation*, 161.
8 Bond, *Report on Bonded Canadian Labor*, 13.
9 Andrew Egan and Deryth Taggart, "Public Perception of the Logging Profession in Maine and Implications for Logger Recruitment," *Northern Journal of Applied Forestry* 26, no. 3 (2009): 93–98.
10 Hillard, "Interview with Don Fontaine."
11 Osborn, *Paper Plantation*, 166.
12 John Kifner, "Maine Woodsmen Embittered," *New York Times*, November 29, 1975, 15.
13 Bill Parenteau, "The Rise of the Small Contractor: A Study of Technological and Structural Change in the Maine Pulpwood Industry" (master's thesis, University of Maine at Orono, 1986), 180.
14 *Great Northern Newsletter for Management Employees*, June 20, 1975, 103–4, Public Affairs Office of Great Northern Paper Company, https://digitalcommons.library.umaine.edu/cgi/viewcontent.cgi?article=1096&context=great_northern.
15 John Kifner, "Maine Woodsmen Embittered," 15.
16 Michael Hillard, "Interview with Claire Bolduc," July 7, 2004, "Maine Woodman's Association Strike of 1975," digitalcommons.usm.maine.edu/strike/.
17 Michael Hillard, "Interview with Jon Falk," July 27, 2004, "Maine Woodman's Association Strike of 1975," digitalcommons.usm.maine.edu/strike/.
18 David Vail, "How to Tell the Forest from the Trees," *Technology and Science* 11 (1989): 355.
19 *Great Northern Newsletter for Management Employees*, December 19, 1980, 185, Public Affairs Office of Great Northern Paper Company, https://digitalcommons.library.umaine.edu/cgi/viewcontent.cgi?article=1103&context=great_northern.
20 *Great Northern Newsletter for Management Employees*, January 30, 1981, 17, Public Affairs Office of Great Northern Paper Company, https://digitalcommons.library.umaine.edu/cgi/viewcontent.cgi?article=1104&context=great_northern.
21 *Great Northern Newsletter for Management Employees*, March 20, 1981, 49, Public Affairs Office of Great Northern Paper Company, https://digitalcommons.library.umaine.edu/cgi/viewcontent.cgi?article=1104&context=great_northern.
22 *Cut and Run*, directed by Richard Searls (Bucksport, ME: Northeast Historic Film, 1980), DVD.
23 Searls, *Cut and Run*; Hillard, "Interview with Claire Bolduc."
24 Searls, *Cut and Run*.
25 Bond, *Report on Bonded Canadian Labor*, 4.
26 *Great Northern Newsletter for Management Employees*, October 17, 1975, 151–52, Public Affairs Office of Great Northern Paper Company, https://digitalcommons.library.umaine.edu/cgi/viewcontent.cgi?article=1096&context=great_northern.
27 *Great Northern Newsletter for Management Employees*, November 15, 1975, 172, "Public Affairs Office of Great Northern Paper Company," https://digitalcommons.library.umaine.edu/cgi/viewcontent.cgi?article=1096&context=great_northern.
28 Hillard, "Interview with Don Fontaine."
29 Hillard, "Interview with Don Fontaine."
30 Michael Hillard, "Interview with Peter Haggarty," August 18, 2004, University of Southern Maine Digital Commons, "Maine Woodman's Association Strike of 1975," digitalcommons.usm.maine.edu/strike/.

31 Michael Hillard, "Interview with Bill Butler," July 27, 2004, University of Southern Maine Digital Commons, "Maine Woodman's Association Strike of 1975," digitalcommons.usm.maine.edu/strike/.
32 "Woodsmen and Paper Industry Wage Bitter Warfare in Maine," *New York Times*, October 19, 1975, 45.
33 Michael Hillard, *Shredding Paper: The Rise and Fall of Maine's Mighty Paper Industry* (Ithaca, NY: ILR Press, 2020), 286.
34 "Woodsmen and Paper Industry."
35 Hillard, "Interview with Don Fontaine."
36 "Woodsmen and Paper Industry."
37 Hillard, "Interview with Bill Butler."
38 Hillard, "Interview with Mel Ames."
39 Hillard, "Interview with Don Fontaine."
40 "Woodsmen and Paper Industry."
41 David Vail, "The Internal Conflict: Contract Logging, Chainsaws, and Clearcuts in Maine Forestry," in *Who Will Save the Forests? Knowledge, Power and Environmental Destruction*, ed. Tariq Banuri and Frederique Apffel-Marglin (London: Zed Books, 1993), 164.
42 Hillard, "Interview with Mel Ames."
43 Vail, "Internal Conflict," 165.
44 Kifner, "Maine Woodsmen Embittered."
45 Bond, *Report on Bonded Canadian Labor*, 1.
46 Searls, *Cut and Run*.
47 Hillard, "Interview with Don Fontaine."
48 Hillard, "Interview with Jon Falk."
49 Hillard, "Interview with Claire Bolduc."
50 Industry News, *Northern Logger*, November 1975.
51 William Siegel, "The Legal Distinction between Employee, Employer, and Independent Contractor as Applied to Collective Bargaining by Woods Workers" (paper presented at the annual meeting, New England Section, Society of American Foresters, Boston, MA, March 10, 1976). Recognizing a quirk in the interpretation of an agricultural worker versus a forest worker, Siegel wrote, "The courts have held that ornamental nursery employees are agricultural workers, but the National Labor Relations Board has held that forest tree nursery employees are not" (2, 3).
52 Hillard, "Interview with Bill Butler."
53 Dan Morgan, "Company 'Slavery' Ignites Maine Woodmen's Union," *Washington Post*, September 6, 1977.
54 Hillard, "Interview with Claire Bolduc."
55 Hillard, "Interview with Jon Falk."
56 Bond, *Report on Bonded Canadian Labor*, 2, 6–7, and 13.
57 George Fowler, "Unions in the Woods? Will It Be Unions or Cooperatives?," *Northern Logger and Timber Processor*, August 1979, 40.
58 Troy Jackson, "Here's Why the 1998 Logging Blockade Still Matters," *Bangor Daily News*, October 25, 2019.
59 Tony Kryzanowski, "Loggers' Blockade at Canada/US Border on Hold—for Now," *Logging and Sawmilling Journal*, December–January 1999.
60 Pan Atlantic Consultants and the Irland Group Forestry Consultants, *Maine Logging Industry and the Bonded Labor Program: An Economic Analysis* (Augusta: Maine Department of Labor, 1999).

61 Because I had the most current and comprehensive data from nearly nine hundred loggers who worked in Maine, I was asked to provide testimony to the committee. Several of my peer-reviewed papers were cited in the report.

62 Retrieved at Maine Legislative Committee to Study New Payment Models for the Logging Industry, p. i, https://legislature.maine.gov/doc/2049.

63 Letter from Maine State representatives Troy Jackson and Bill Smith, dated August 6, 2004, Maine House Democrats, "Northern Maine Legislators Urge Maine Congressional Delegation to Keep Bonded Worker Cap in Place," http://www.maine.gov/tools/whatsnew/index.php?topic=HouseDems+News&id=3161&v=Article.

64 Daniel Dructor, "Importing Loggers?," *Bangor Daily News*, July 20, 2004.

65 Barry Newman, "Chain-Saw Reaction: Rules Restricting Canadian Loggers Backfire on Maine—Visa Shortage Cuts Labor Pool and Reduces Production; Higher Prices Imperil Jobs," *Wall Street Journal*, April 25, 2005.

66 Quotations from focused discussion with author.

67 Jack Cashman, *Governor's Council on the Sustainability of the Forest Products Industry*, final report (Augusta: Maine Department of Economic and Community Development, 2005), 38.

68 Michael Hillard, "Interview with Lucien Duchaine," June 27, 2006, University of Southern Maine Digital Commons, "Maine Woodman's Association Strike of 1975," digitalcommons.usm.maine.edu/strike/.

69 Nick Sambides, "Politicians Say Irving Work Halt 'Blackmail,'" *Bangor Daily News*, June 1, 2009.

70 Kevin Miller, "State Calls for Logging Inquiries," *Bangor Daily News*, July 23, 2009.

71 Miller, "State Calls for Logging Inquiries."

72 "Logging Feud Escalates over Foreign Hires," Mainebiz.biz, August 24, 2009, https://www.mainebiz.biz/article/logging-feud-escalates-over-foreign-hires.

73 Kevin Miller, "Loggers Threaten Blockade," *Bangor Daily News*, July 28, 2009.

74 Kevin Miller, "Logging Outfits Punished for Labor Violations," *Bangor Daily News*, November 13, 2009.

75 Julia Bayly, "Woodsmen Forum Targets Logging Regulations," *Bangor Daily News*, May 7, 2010. The 2010 Act to Protect Maine Workers can be found online at https://www.mainelegislature.org/legis/bills/bills_124th/chappdfs/PUBLIC637.pdf.

76 Julia Bayly, "Logging Forum Scheduled in Fort Kent," *Bangor Daily News*, May 3, 2010.

77 Julia Bayly, "Woodsmen Forum Targets Logging Regulations."

78 Kevin Miller, "Logging Companies Cited for Foreign Equipment," *Bangor Daily News*, October 27, 2010.

79 Steve Mistler, "GOP Takes Heat for Opposing Bills Dealing with Canadian Loggers in Maine Woods," *Bangor Daily News*, April 27, 2011.

80 John Keefe, "Feds Allow Canadian Loggers to Threaten Maine Jobs, Senate President Says in Complaint," *Bangor Daily News*, September 18, 2020.

81 "Study of the Statewide Market for Forest Products Harvesting and Hauling Services, February 2009, 3. The study can be found at https://www.maine.gov/ag/dynld/documents/Final%20logging%20study.pdf.

82 State of Maine, "Study of the Statewide Market," 6.

83 State of Maine, "Study of the Statewide Market," 18.

84 State of Maine, "Study of the Statewide Market," 19.

85 Kevin Miller, "Border Tensions: Maine Loggers Contend Canadian Companies Skirting Labor Laws," *Bangor Daily News*, August 6, 2009.

86 Bayly, "Logging Forum Scheduled in Fort Kent."

87 Andrew Egan and Deryth Taggart, "Who Will Log? Occupational Choice and Prestige in Northern New England," *Journal of Forestry* 102, no. 1 (2004): 20–25; Andrew Egan and Deryth Taggart, "Who Will Log in Maine's North Woods? A Cross-Cultural Study of Occupational Choice and Prestige," *Northern Journal of Applied Forestry* 21, no. 4 (2004): 200–208.

88 Staff, "Report: Shortage of Loggers and Truckers Hinders Growth of $8.5B Forest Products Industry," Mainebiz.biz, March 15, 2019, https://www.mainebiz.biz/article/report-shortage-of-loggers-and-truckers-hinders-growth-of-85b-forest-products-industry.

89 "An Act to Expand Application of the Maine Agricultural Marketing and Bargaining Act of 1973 to Harvesters and Haulers of Forest Products," 2019, https://legislature.maine.gov/legis/bills/bills_129th/chapters/PUBLIC248.asp.

90 Mal Leary, "Senate President Troy Jackson's Bill Would Allow Loggers to Unionize," April 29, 2019, mainepublic.org, https://www.mainepublic.org/politics/2019-04-29/senate-president-troy-jacksons-bill-would-allow-loggers-to-unionize.

91 Hillary Lister, "Loggers and Truckers in Northern Maine Held Out for Three Weeks in a Strike against Canadian Land-Owning Giant Irving," February 27, 2004, maineindymedia.org/newswire/display/960/index.php (no longer available).

92 Maureen Milliken, "Loggers Cooperative, Formed after New Law, Partners with Quebec Company," Mainebiz.biz, November 1, 2019, https://www.mainebiz.biz/article/loggers-cooperative-formed-after-new-law-partners-with-quebec-company.

CHAPTER SIX: WHO WILL LOG?

1 The title of chapter six derives from the article by Andrew Egan and Deryth Taggart, "Who Will Log? Occupational Choice and Prestige in Northern New England," *Journal of Forestry* 102, no. 1 (2004): 20–25.

2 L. P. Bissell, "Can the Woods Labor Problem be Solved?," *Northern Logger and Timber Processor*, 1968, 62.

3 J. Maines, "Appears That 1968 Will Be a Good Year," *Northern Logger and Timber Processor*, January 1968, 15.

4 "Imported Workers Line Up for Jobs Americans Shun," *New York Times*, August 25, 1967; Great Northern Newsletter for Management Employees, 1967, https://digitalcommons.library.umaine.edu/cgi/viewcontent.cgi?article=1094&context=great_northern; Great Northern Newsletter for Management Employees, 1968, https://digitalcommons.library.umaine.edu/cgi/viewcontent.cgi?article=1095&context=great_northern.

5 Bissell, "Can the Woods Labor Problem be Solved?," 62.

6 From discussion led by author.

7 Peter MacDonald and Michael Clow, "Just One Damn Machine after Another—Technological Innovation and the Industrialization of Tree Harvesting Systems," *Technology in Society* 21 (1999): 323–44 (quotations on 328–29).

8 Egan and Taggart, "Who Will Log?," 24.

9 Camille Legendre, *Le Travailleur Forestier Québécois* (Québec: Presse de l'Université du Québec, 2005), 170. Author's translation.

10 Quotations from focused discussion between eleven loggers and author, Québec, 2005.

11 Louis-Pierre Bélanger-Ducharme, Luc LeBel, and Andrew Egan, "Forest Entrepreneurs in Québec: Current and Future Challenges" (paper presented at Council

on Forest Engineering meeting, Charleston, SC, June 22–25, 2008, https://cofe.org/pdfs/COFE_2008.pdf.

12 Egan and Taggart, "Who Will Log?," 22–23.

13 Andrew Egan and Deryth Taggart, "Who Will Log in Maine's North Woods? A Cross-Cultural Study of Occupational Choice and Prestige," *Northern Journal of Applied Forestry* 21, no. 4 (2004): 200–208. To identify loggers in the region, the authors used mailing lists of loggers in Maine's border counties with Québec that they derived from two sources: lists of both English- and French-speaking loggers who had been trained by the Certified Logging Professionals (CLP) Program and a list of loggers who had filed a harvest notification with the state of Maine in 2000. A version of the survey in French was mailed to French-speaking loggers who worked in the five-county Maine-Québec border region. (CLP certification is required of all loggers who sell wood to mills participating in the Sustainable Forestry Initiative [SFI], a forest products industry–sponsored initiative designed to set standards for and monitor logging quality. Because the forests of northern Maine are dominated by industry ownerships and wood-consuming mills that participate in SFI, CLP certification is required of virtually all loggers who work in the study region.) Entries on the CLP list of French-speaking loggers were all from the Province of Québec. Multiple survey mailings—a cover letter and survey, followed by a reminder postcard, and finally a second cover letter and survey—were sent to 2,318 loggers on the combined lists. Loggers whose names appeared on more than one list were sent only one sequence of mailings. Overall, 626 loggers responded to the survey. Of these, 442 were Maine resident loggers who worked in the state's border counties with Québec, and 184 were Québec resident loggers who worked in these counties.

14 Egan and Taggart, "Who Will Log?," 22–24; Egan and Taggart, "Who Will Log in Maine's North Woods?," 205.

15 State of Maine 121st Legislature First Regular Session, *Final Report of the Committee to Study New Payment Models for the Logging Industry* (Augusta: Office of Policy and Legal Analysis, Maine Legislature, 2004), https://legislature.maine.gov/doc/2049, i–ii.

16 State of Maine, "Study of the Statewide Market for Forest Products Harvesting and Hauling Services," 2009, www.maine.gov/ag/dtnld/documents/Final%20logging%20study.pdf (no longer available).

17 Deryth Taggart and Andrew Egan, "Logging across Borders and Cultures: An Example in Northern Maine," *Forest Products Journal* 61, no. 7 (2012): 561–69.

18 James Wesley McNutt, "*Labor Relations in the Forest*," Weyerhaeuser Lecture Series, Faculty of Forestry and Landscape Architecture, University of Toronto, November 1978, 7.

19 H. Shirley, "This Business of Logging," *Northern Logger and Timber Processor*, June 1954, 11.

20 David Vail, "The Internal Conflict: Contract Logging, Chainsaws, and Clearcuts in Maine Forestry," in *Who Will Save the Forests? Knowledge, Power and Environmental Destruction*, ed. Tariq Banuri and Frederique Apffel-Marglin (London: Zed Books, 1993), 156.

21 Study cited in Egan and Taggart, "Who Will Log?," 20–25; Egan and Taggart, "Who Will Log in Maine's North Woods?," 200–208; Elliot A. Krause, *The Sociology of Occupation* (Boston: Little, Brown, 1971), 393; Charles B. Nam and Mary G. Powers, *The Socioeconomic Approach to Status Measurement* (Houston: Cap and Gown Press, 1983), 147.

22 Albeno Garbin and Frederick Bates, "Occupational Prestige: An Empirical Study of Its Correlates," *Social Forces* 40 (1961): 131–36.

23 Donald Treiman, *Occupational Prestige in Comparative Perspective* (New York: Academic Press, 1977), 514.

24 Ronald M. Pavalko, *Sociology of Occupations and Professions* (Itasca, IL: F. E. Peacock Publishers, 1971), 112; Ronald M. Pavalko, *Sociology of Occupations and Professions*, 2nd ed. (Itasca, IL: F. E. Peacock Publishers, 1988), 355.

25 C. Ross Silversides, *Broadaxe to Flying Shear: The Mechanization of Forest Harvesting East of the Rockies* (Ottawa: National Museum of Science and Technology, 1997), 174.

26 Donald Treiman, "Problem of Concept and Measurement in the Comparative Study of Occupational Mobility," *Social Science Research* 4 (1975): 183–230.

27 Nam and Powers, *Socioeconomic Approach to Status Measurement*, 147.

28 Les Krantz, *Jobs Rated Almanac: The Best and Worst Jobs* (New York: St. Martin's Griffin, 2000); Treiman, *Occupational Prestige in Comparative Perspective*, 514.

29 Andrew Egan, *Adirondack Hard Times: Evolution of a Rich Man's Paradise* (Charleston, SC: History Press, 2021).

30 Legendre, *Le Travailleur Forestier Québécois*, 157, 209. Author's translation.

31 Egan and Taggart, "Who Will Log in Maine's North Woods?," 205–6.

32 Matthew S. Carroll, *Community and the Northwestern Logger: Continuities and Changes in the Era of the Spotted Owl* (Boulder, CO: Westview Press, 1995), 178.

33 Theresa A. Satterfield, "Pawns, Victims, or Heroes: The Negotiation of Stigma and the Plight of Oregon's Loggers," *Journal of Social Issues* 52, no. 1 (1996): 71–83.

34 Egan and Taggart, "Who Will Log in Maine's North Woods?," 205.

35 Andrew Egan and Deryth Taggart, "Public Perception of the Logging Profession in Maine and Implications for Logger Recruitment," *Northern Journal of Applied Forestry* 26, no. 3 (2009): 95–96.

36 Egan and Taggart, "Public Perception of the Logging Profession in Maine," 96–97, 98.

37 Deryth Taggart and Andrew Egan, "Occupational Choice among Loggers in Maine's Northern Forest and Southern Counties," *Adirondack Journal of Environmental Studies* 9, no. 2 (Fall/Winter 2002): 31–36.

38 Taggart and Egan, "Occupational Choice among Loggers," 36.

39 Egan and Taggart, "Public Perception of the Logging Profession," 93–98; see esp. 96–97. Analysis indicated a significant difference in these two populations of Maine residents on the questions of whether they would encourage a son or daughter to become a logger and logging's social prestige.

40 Differences between northern and southern Maine residents were not statistically significant on these questions.

41 Egan and Taggart, "Public Perception of the Logging Profession," 98.

42 Not all differences related to "place" (northern versus southern Maine) in "Public Perception of the Logging Profession" were statistically significant. Those that were included responses to the following questions or statements: Would you encourage a son or daughter to become a logger?; I did not consider becoming a logger because logging has little social prestige; and I did not become a logger because there are better jobs available in the area in which I live.

43 Leo Thibodeau, "Effective Recruiting and Training of Woods Workers and Loggers," *Northern Logger and Timber Processor*, May 1965, 13. Thibodeau's paper was

presented at the Open Industry Session of the American Pulpwood Association's Annual Meeting in New York on February 24, 1965.

44 Thibodeau, "Effective Recruiting and Training," 13.

45 Thibodeau, 31.

46 Thibodeau, 31.

47 George Carlisle, "Comments on Mr. Leo Thibodeau's Paper," *Northern Logger and Timber Processor*, May 1965, 13.

48 Hardin R. Glascock, "Comments from the West on Effective Recruiting and Training of Woodsworkers and Loggers," *Northern Logger and Timber Processor*, May 1965, 14. Glascock's paper was presented at the Annual Meeting of the American Pulpwood Association in New York on February 24, 1965.

49 Richard Donovan and Elizabeth Swain, *Maine Woods Labor Supply, 1984–85* (Augusta: Maine Department of Conservation, 1986), 20.

50 *Cut and Run*, directed by Richard Searls (Bucksport, ME: Northeast Historic Film, 1980), DVD.

51 James E. Granskog and Robert S. Manthy, *Maintaining Michigan's Woods Labor Supply*, Agricultural Experiment Station Research Bulletin 28 (Michigan State University, 1970), 36.

52 Donovan and Swain, *Maine Woods Labor Supply*, 6.

53 Dale Greene, Ben Jackson, and David Woodruff, "Characteristics of Logging Contractors and Their Employees in Georgia," *Forest Products Journal* 48, no. 1 (1998): 47–53.

54 Andrew Egan, Curt Hassler, and Shawn Grushecky, "Are Logger Certification and Training Worthwhile? A View from West Virginia's Logging Community," *Forest Products Journal* 47, no. 7/8 (1997): 46–50, https://www.thefreelibrary.com/Training+preferences+and+attitudes+among+loggers+in+northern+New . . . -a0131043471.

55 Unpublished data from the 2006 Northeast United States Professional Logger Survey, conducted by Egan while on the faculty of Laval University. Maine logger responses were disaggregated from the broader logger respondent population. Responses are from Question 21: "In your opinion, what could be done to attract people to the logging profession?"

56 "Logging Adds Over $600 Million to Maine Economy, but the Industry Can't Find Workers," Mainebiz.biz, March 6, 2020, https://www.mainebiz.biz/article/logging-adds-over-600m-to-maine-economy-but-the-industry-cant-find-workers.

57 Maine Department of Labor Workforce Outlook to 2028, https://www.maine.gov/labor/cwri/outlook.html.

58 "Logging Adds Over $600 Million to Maine Economy, but the Industry Can't Find Workers," Mainebiz.biz, March 6, 2020; Nina Mahaleris, "This Program That Trains New Loggers Shows Promise for Maine's Industry," *Bangor Daily News*, March 5, 2020.

59 Bill Trotter, "Logging Industry Works on Attracting Recruits," *Bangor Daily News*, January 7, 2006.

60 Egan and Taggart "Who Will Log," 22; Staff, "Report: Shortage of Loggers and Truckers Hinders Growth of $8.5B Forest Products Industry," Mainebiz.biz, March 15, 2019, https://www.mainebiz.biz/article/report-shortage-of-loggers-and-truckers-hinders-growth-of-85b-forest-products-industry.

61 Egan and Taggart, "Who Will Log," 22.

62 Peter Blombäck and Peter Poschen, "Decent Work in Forestry? Enhancing Forestry Work and Forest-Based Livelihoods," in Congress Proceedings, p. 9, XII World Forestry Congress, September 21–28, 2003, Québec City.

CHAPTER SEVEN: CRISIS OR CROSSROADS?

1 Nick Sambides, "Roxanne Quimby Calls Maine a 'Welfare State,'" *Bangor Daily News*, October 7, 2011. Regarding her observation about obesity, Quimby may have been influenced by an article that appeared in the *Bangor Daily News* three months earlier stating that "Maine retains the dubious distinction of being the most obese of the New England states." See Meg Haskell, "Mainers Most Obese in New England; 27th Fattest in Nation, Up from 35th in 2009," *Bangor Daily News*, July 7, 2011.

2 Sambides, "Maine a 'Welfare State.'"

3 Andrew Egan, *Adirondack Hard Times: Evolution of a Rich Man's Paradise* (Charleston, SC: History Press, 2021), 102.

4 International Snowmobile Manufacturers Association, https://www.snowmobile.org/snowmobiling-statistics-and-facts.html.

5 "2011 Snowmobile Owners Survey: Executive Summary," https://nysnowmobiler.com/wp-content/uploads/2015/11/2011-snowmobile-economic-impact-study.pdf.

6 Hannah Catlin, "Pandemic-Altered Snowmobile Season Puts the Squeeze on St. John Valley Hotels," *Bangor Daily News*, February 27, 2021.

7 Saint Regis Canoe Area Unit Management Plan, New York State Department of Environmental Conservation, June 2006, https://www.dec.ny.gov/docs/lands_forests_pdf/srcafinal.pdf.

8 "State Should Not Buy Whitney Park," editorial, *Lake Placid News*, August 13, 2020.

9 *The economic importance and wood flows from New York's Forests, 2007*, report of the North East State Foresters Association, August 2007, https://www.nefainfo.org/uploads/2/7/4/5/27453461/nefaeconomicimportny.pdf, 2–7.

10 Brian Mann, "Adirondack Tourism Lifts Some Boats," Adirondack Almanack, July 8, 2014, https://www.adirondackalmanack.com/2014/07/brian-mann-adirondack-tourism-raises-some-boats.html.

11 Kerri Arsenault, *Mill Town: Reckoning with What Remains* (New York: St. Martin's, 2020), 354.

12 Jonathan Wood, "A Sustainable Resource for a Sustainable Rural Economy," in *The Future of the Northern Forest*, ed. Christopher McGrory Klyza and Stephen Trombulak (Hanover, NH: Middlebury College Press, 1994), 169–70.

13 Eric H. Wharton, Richard H. Widmann, Carol L. Alerich, Charles J. Barnett, Andrew J. Lister, Tonya W. Lister, Don Smith, and Fred Borman, *The Forests of Connecticut*, Northeastern Research Bulletin NE-160 (Newtown Square, PA: United States Department of Agriculture, Forest Service, 2004), 36.

14 Susan Campbell and David Kittredge, "Woodscaping Crew for Small Woodlot Management in Southeastern Massachusetts," *Northern Journal of Applied Forestry* 9 (1992): 116–18.

15 Sarah Schweitzer, "For Loggers, a Way of Life Teeters: Distressed Industry Hit Hard by New Rules on Mass. Forests," *Boston Globe*, April 26, 2010.

16 "Massachusetts Expands Public Lands Ban," Late Industry News, *Northern Logger and Timber Processor*, May 2010, 2.

17 David Smith, "Adapting Forestry to Megalopolitan Southern New England," *Journal of Forestry* 67, no. 7 (1969): 372–77, (quotation on 372).

18 Campbell and Kittredge, "Woodscaping Crew," 116–18.

19 David Kittredge, Michael Mauri, and Edward McGuire, "Decreasing Woodlot Size and the Future of Timber Sales in Massachusetts: When Is an Operation Too Small?," *Northern Journal of Applied Forestry* 13, no. 2 (1996): 96–101 (data are on 96).

20 Andrew Egan, "Characteristics of and Challenges Faced by Logging Businesses in Southern New England," *Northern Journal of Applied Forestry* 28, no. 4 (2011): 183.

21 "MassWoods," UMass Amherst, https://masswoods.org/massachusetts-forests.

22 Andrew Egan and Louis Morin, "Challenges to Sawmill Businesses in New England and New York State: A Survey of Wood Procurement Managers," *Journal of Forestry* 108, no. 8 (2010): 408–12.

23 From a focused discussion with seven Connecticut loggers on June 23, 2005.

24 Egan, "Characteristics of and Challenges Faced by Logging Businesses," 183–84.

25 Derived from data reported in Deryth Taggart and Andrew Egan, "Occupational Choice among Loggers in Maine's Northern Forest and Southern Counties," *Adirondack Journal of Environmental Studies* 9, no. 2 (Fall/Winter 2002): 31–36, and Egan, "Characteristics of and Challenges Faced by Logging Businesses," 194–98. Results should be interpreted with some caution, because statistical tests of independence were not conducted on the combined data from which these results were derived.

26 Andrew Egan, Deryth Taggart and Isaac Annis, "Effects of Population Pressures on Wood Procurement and Logging Opportunities in Northern New England," *Northern Journal of Applied Forestry* 24, no. 2 (2007): 85–90.

27 William Fulton, Rolf Pendall, Mai Nguyen, and Alicia Harrison, *Who Sprawls Most? How Growth Patterns Differ across the U.S.* (Washington, D.C.: Brookings Institution, Center on Urban and Metropolitan Policy, July 2001), 24.

28 Lloyd C. Irland, "This Evergreen Empire: Maine's Forest Resources and Industries in a New Century," Appendix H, Blaine House Conference on Maine's Natural Resource-Based Industry: Charting a New Course, November 17, 2003.

29 Kenneth Laustsen, Douglas Griffith, and James Steinman, *Fourth Annual Inventory Report on Maine's Forests* (Augusta: Department of Conservation, Maine Forest Service, 2003), 55; George L. McCaskill, William H. McWilliams, Charles J. Barnett, Brett J. Butler, Mark A. Hatfield, Cassandra M. Kurtz, Randall S. Morin, W. Keith Moser, Charles H. Perry, and Christopher W. Woodall, *Maine's Forests 2008*, Resource Bulletin NRS-48, p. 8, https://www.nrs.fs.fed.us/pubs/rb/rb_nrs48.pdf.

30 David R. Foster, *Wildlands and Woodlands: A Vision for the New England Landscape* (Cambridge, MA: Harvard University Press, 2010).

31 The term *terminal harvest* is defined by Sarah Thorne and Dan Sundquist in *New Hampshire's Vanishing Forests: Conversion, Fragmentation, and Parcelization of Forests in the Granite State*, Report of the New Hampshire Forest Land Base Study, p. 1, April 2001, https://forestsociety.org/sites/default/files/New%20

Hampshire%27s%20Vanishing%20Forests_%20-%20Society%20for%20the%20Protection%20of%20. . . . pdf. Terminal harvesting is different from so-called liquidation harvesting, described by the Maine Forest Service as "the purchase of timberland followed by a harvest that removes most or all commercial value in standing timber, without regard for long-term management principles, and the subsequent sale or attempted resale of the harvested land within 5 years" (https://www.maine.gov/dacf/mfs/publications/rules_and_regs/chap_23_rules.pdf).

32 Egan, Taggart, and Annis, "Effects of Population Pressures," 85–90.

33 I was among the relatively small number of logging employees in New Hampshire who rode the wave of biomass harvesting in the early 1980s at a time when biomass markets were more stable, mechanized logging equipment less expensive, and contracts for the delivery of whole-tree chips more reliable. Hired as a chopper, then a skidder and feller-buncher operator, I worked for several years for a logging contractor who grew his business from a relatively small, conventional hand crew, comprising two cable skidders, several chain saws, and a triaxle log truck, to one that included two feller-bunchers, two grapple skidders, and a whole-tree chipper. Biomass contracts for whole-tree chips were with S. D. Warren in Westbrook, Maine, which was purchased by Scott Paper in 1967. Scott sold the mill in 1994 to SAPPI Limited, a South African paper company. Employment at the mill dropped from some three thousand workers at its peak in the 1900s to about three hundred workers in the 2000s.

34 Eric Kingsley, "Tough Time to Try to Sell Biomass Fuel," *Northern Logger and Timber Processor*, November 2016, 16–18 (quotation on 18).

35 "As Loggers Get Stiffed, Maine Learns a Lesson about Propping Up Struggling Industries," *Bangor Daily News*, March 23, 2017.

36 Mary S. Booth, "Maine's Bioenergy at the Crossroads," Partnership for Policy Integrity, April 18, 2017, DigitalCommons@UMaine, https://digitalcommons.library.umaine.edu/maine_env_organizations/229. Readers interested in an alternative treatment of the bioenergy sector in Maine should consider reading Booth's full report.

37 James McCarthy, "Loggers Suffer One-Two Punch with Mill and Biomass Plant Closures," Mainbiz.biz, April 4, 2017.

38 "Biomass Subsidies, Long-Term Energy Contracts Aren't Good Ways to Help Maine loggers," *Bangor Daily News*, April 6, 2016.

39 The type of certification used to determine the "green" supply network for woody biomass can matter politically, since, deserved or not, forest industry–sponsored certifications have been the object of some criticism for their perceived conflict of interest. For example, the director (or "executive director" or "coordinator," depending on which website one visits) of Maine's Sustainable Forestry Initiative (SFI) is listed as a staff member of the Maine Forest Products Council—again, "the voice of Maine's forest economy"—on that organization's website. Maine's SFI homepage doesn't mention this affiliation. Does Maine's SFI lead report to the executive director of the Maine Forest Products Council? Again according to its website, Maine SFI is apparently "run by a State Implementation Committee (SIC), comprised of various stakeholder groups." Yet, in 2022 the link to that committee ran the message, "Apologies, but the page you requested could not be found," leaving one with no idea of who populates that committee, unless one digs further to find the "SIC annual reports," which did not appear to be

published annually. At a minimum, the messaging was sloppy and confusing and the optics potentially disquieting to potential consumers of SFI's forest certification services, reflecting poorly on an organization that may otherwise have merit. For the Maine Forest Products Council, "Staff," see https://maineforest.org/about-us/staff/; for SFI Maine, "Who Runs SFI," see https://sfimaine.org/new-page/.

40 Eric Johnson, "Biomass Markets under Pressure," *Northern Logger and Timber Processor*, 2015, 22–24.

41 Forest Opportunity Roadmap—FOR/Maine, https://formaine.org/; Forest Opportunity Roadmap/Maine Report, September 2018, https://formaine.org/wp-content/uploads/2020/09/FORMaine_Report_DL_041119.pdf.

42 "What Happened to Maine's Pulp and Paper Industry?," radio report, WGME, September 8, 2017, https://wgme.com/news/in-depth/what-happened-to-maines-pulp-and-paper-industry.

43 Maine's Pulp and Paper Industry by the Numbers, radio report, WGME, September 8, 2017, https://wgme.com/news/in-depth/maines-pulp-and-paper-industry-by-the-numbers.

44 Arsenault, *Mill Town*, 176.

45 George Wuerthner, "Why Is Logging Dying? Blame the Market," *High Country News*, June 15, 2016.

46 Lloyd Irland, "When the Mill Goes Quiet: Maine's Paper Industry, 1990–2016," *Maine History* 54, no. 1 (2021): 5–26.

47 Jack Cashman, chair, Governor's Council on the Sustainability of the Forest Products Industry, *Final Report*, March 2005, https://www.inrsllc.com/download/SustainabilityForestProductsIndustry.pdf.

48 Kevin Miller, "Border Tensions: Maine Loggers Contend Canadian Companies Skirting Labor Laws," *Bangor Daily News*, January 27, 2009, https://bangordailynews.com/2009/08/06/news/border-tensions-maine-loggers-contend-canadian-companies-skirting-labor-laws/.

49 "Maine's Forest Economy," 2013, p. 4, https://maineforest.org/wp-content/uploads/2013/09/Maines-forest-economy.pdf; Patrick Strauch, "Maine Voices: Resilient Forest Economy Evolves," Pressherald.com, November 10, 2013, https://www.pressherald.com/2013/11/10/maine_voices__resilient_forest_economy_evolves_/.

50 Darren Fishell, "Verso Mill in Bucksport to Close by Year's End, 570 Employees to Lose Jobs," *Bangor Daily News*, October 1, 2014.

51 Nok-Noi Ricker, "Old Town Mill to Close," *Bangor Daily News*, September 29, 2015; Nick Sambides, "Massive Fire Burns Former Lincoln Paper Mill," *Bangor Daily News*, November 15, 2017.

52 Darren Fishell, "Madison Paper Mill Shuts Down," *Bangor Daily News*, May 25, 2016.

53 Eric Kingsley, "The State of Maine's Pulp and Paper Industry," *Northern Logger and Timber Processor*, February 2016, 14–15, 26–27 (quotations on 14).

54 James McCarthy, "Maine Pulp and Paper Association Folds," Mainebiz.biz, January 16, 2017.

55 Dawn Gagnon, "Maine Pulp and Paper Trade Group Dissolves after Half a Century," *Bangor Daily News*, January 15, 2017.

56 Eric Kingsley, "The State of the N.E. Regional Pulp Market," *Northern Logger and Timber Processor*, February 2017, 10–11.

57 Nick Sambides, "Sappi to Cut 75 Jobs at Westbrook Paper Mill," *Bangor Daily News*, July 9, 2020; Lori Valigra, "Jay Mill to Lay Off 59 Workers after Explosion," *Bangor Daily News*, July 9, 2020.

58 Matthew Stone, "Here's Where Redevelopment Stands at 6 Maine Paper Mills That Have Closed since 2008," *Bangor Daily News*, September 26, 2019.

59 James McCarthy, "Loggers Suffer One-Two Punch with Mill and Biomass Plant Closures," Mainebiz.biz, April 4, 2016.

60 Bill Donahue, "The Paper Mill Was Everything to the People of Jay, Maine. Then It Exploded," *Boston Globe*, August 12, 2020.

61 Quoted in "Maine's Logging Industry Confronts Pulpwood 'Crisis,'" *Bangor Daily News*, December 31, 2015.

62 Eric Kingsley, "The State of Pulpwood in the State of Maine," *Northern Logger and Timber Processor*, November 2015, 22–25.

63 Irland, "When the Mill Goes Quiet," 5–26 (quotation on 19).

64 Arsenault, *Mill Town*, 258.

65 Late Industry News, *The Northern Logger and Timber Processor*, January 2016.

66 Late Industry News, *The Northern Logger and Timber Processor*, December 2014.

67 Donahue, "The Paper Mill Was Everything."

68 "Maine's Logging Economy—A Study of the Impact of Professional Logging in Maine," Professional Logging Contractors of Maine, 2014, http://maineloggers.com/new/wp-content/uploads/2016/09/Logging-Economic-Impact-Study-2014-brochure-FINAL-web-version.pdf (no longer available).

69 "The State of Canada's Forests," Natural Resources Canada, 2007, http://cfs.nrcan.gc.ca:80/sof/sof05/profilesQC_e.html (no longer available); Réginald Harvey, "Il Faut Sauver la Foret Québécoise," *Le Devoir*, March 1, 2006; Kevin Dougherty, "Forest Cut Is the Deepest," *The Gazette* (Montreal), December 9, 2006; éric Desrosiers, "Kruger "Consternée" par le Rapport du Forestier en Chef," *Le Devoir*, December 14, 2006; Sophie Cousineau, "Forestry's Elusive Quest for Innovation," *Globe and Mail*, November 19, 2013.

70 Norman Delisle, "Forestry Aid Package Paltry, Québec Union Says," *The Gazette* (Montreal), August 22, 2006, B3.

71 Barrie McKenna, "U.S. Warns Ottawa about Lumber Bailout," *Globe and Mail*, February 15, 2006.

72 Roger Blackburn, "La forêt intéresse peu les jeunes," *Le Quotidien* (Saguenay, Québec), February 16, 2006.

73 C. Drolet, "La foresterie, un bon choix de carrière?," *Les Affaires* (Montreal), February 18, 2006.

74 Andrew Egan and Deryth Taggart, "Who Will Log in Maine's North Woods? A Cross-Cultural Study of Occupational Choice and Prestige," *Northern Journal of Applied Forestry* 21, no. 4 (2004): 204.

75 Duncan R. Campbell and Edward B. Power, "Manpower Implications of Prospective Technological Changes in the Eastern Canadian Pulpwood Logging Industry," Canada Department of Manpower and Immigration, Research Branch. Program Development Service, research monograph no. 1 (Ottawa: Queen's Printer, 1966), xii.

76 Andrew Egan and Deryth Taggart, "Who Will Log in Maine's North Woods?," 200–208.

77 Jack Cashman, *Governor's Council on the Sustainability of the Forest Products Industry*, final report (Augusta: Maine Department of Economic and Community Development, 2005), 38.

78 Lloyd Irland, "Here's My Advice for How Maine Can Take On the Challenges Facing Forest Industries," *Bangor Daily News*, August 24, 2016.

79 Charles S. Colgan, "The Maine Economy: Yesterday, Today, and Tomorrow," background paper prepared for the Brookings Institution Metropolitan Policy Program, Brookings Institution, October 2006, https://www.brookings.edu/wp-content/uploads/2016/07/economy.pdf.

80 *Maine Future Forest Economy Project: Current Conditions and Factors Influencing the Future of Maine's Forest Products Industry*, 2006, may be found at https://digitalmaine.com/do/search/?q=maine%20future%20forest%20economy%20project&start=0&context=3363941&facet.

81 Outcomes of this research can be found in the following published papers: Taggart and Egan, "Occupational Choice among Loggers," 31–36; Andrew Egan and Deryth Taggart, "Who Will Log? Occupational Choice and Prestige in Northern New England," *Journal of Forestry* 102, no. 1 (2004): 20–25; Egan and Taggart, "Who Will Log in Maine's North Woods?," 200–208; Andrew Egan, Isaac Annis, W. Dale Greene, Cornelis de Hoop, and Jefferson Mayo, "Unused Logging Capacity in Northern New England," *International Journal of Forest Engineering* 17, no. 1 (2006): 31–38.

82 S. Libby, "An Assessment of Regional Variation in the Factors Affecting the Mill-Delivered Price of Softwood Sawlogs and Pulpwood in Nine States" (master's thesis, College of Natural Science, Forestry, and Agriculture, University of Maine, Orono, 2000), 131.

83 W. Dale Greene, Jefferson Mayo, Cornelis de Hoop, and Andrew Egan, "Causes and Costs of Unused Logging Production Capacity in the Southern States and Maine," *Forest Products Journal* 54, no. 5 (2004): 29–37; Cornelis de Hoop, Andrew Egan, W. Dale Greene, and Jefferson Mayo, "Are 'Preferred Supplier' Contractors Representative of the Logging Business Community? A Survey Analysis," in *Business Research Yearbook 21* (Saline, MI: International Academy of Business Disciplines, 2004): 230–34; Egan, Annis, Greene, de Hoop, and Mayo, "Unused Logging Capacity in Northern New England," 31–38.

84 USDA Census of Agriculture, Maine, 2017, https://www.nass.usda.gov/Publications/AgCensus/2017/Full_Report/Volume_1,_Chapter_1_State_Level/Maine/.

85 According to a report in the *Portland Press Herald*, through a grant to Northeastern University in Boston, Maine native and tech entrepreneur David Roux funded a graduate school and research center in Portland to put Maine "at the forefront of the advanced technology economy." The Silicon Valley millionaire located a $100 million research institute in the state's largest city to help "transform and support an economy that doesn't fully participate in the modern, tech-led innovation economy." Graduate programs include the fields of artificial intelligence, computer and data science, digital engineering, and life sciences. Forestry and wood science were not specifically mentioned. The project is called the Roux Institute at Northeastern University in Maine. See Peter McGuire, "Maine Native Puts Up $1 Million to Open Northeastern Research Center in Portland," *Portland Press Herald*, January 27, 2020.

86 Colgan, "The Maine Economy," 2, 26, 30.

87 Christopher Burns, "What Stands Out about LePage's Approach to the Economy: A Focus on Industries of the Past," *Bangor Daily News*, June 18, 2016.

88 Burns, "LePage's Approach to the Economy." Burns references the 2001 work of Jeffrey Sachs and Andrew Warner, "The Curse of Natural Resources," *European Economic Review* 45, nos. 4–5 (2001): 827–38. Sachs and Warner asserted that "resource-abundant countries tended to be high-priced economies and, perhaps as a consequence, these countries tended to miss-out on export-led growth."

89 See www.census.gov/quickfacts/ME.

90 James Myall, "A New Great Depression in Rural Maine," Maine Center for Economic Policy, July 31, 2017, https://www.mecep.org/maines-economy/a-new-great-depression-in-rural-maine-2/.

91 Eduardo Porter, "A $100 Million Bet That Vacationland Can Be a Tech Hub, Too," *New York Times*, January 30, 2020.

92 The Maine Economic Development Strategy 2020–29 may be found at https://www.maine.gov/decd/strategic-plan.

93 Information on the Governor's Economic Recovery Committee in Maine may be found at https://www.maine.gov/future/initiatives/economy/economic-recovery-committee.

94 Avern Danforth, "Nostalgia isn't an Economic Development Plan for the Katahdin Region," *Bangor Daily News*, June 22, 2015.

95 A. Elliot, "Labor Shortage behind State's Lack of Lumber," *Bangor Daily News*, December 27, 2003.

INDEX

Page numbers marked with *f* or *t* indicate figures and tables, respectively.